Sustainability and Communities of Place

" The End of Poverty ?"

Youtube   Vimeo.

Studies in Environmental Anthropology and Ethnobiology

General Editor: **Roy Ellen**, FBA
Professor of Anthropology, University of Kent at Canterbury

Interest in environmental anthropology has grown steadily in recent years, reflecting national and international concern about the environment and developing research priorities. This major new international series, which continues a series first published by Harwood and Routledge, is a vehicle for publishing up-to-date monographs and edited works on particular issues, themes, places or peoples which focus on the interrelationship between society, culture and environment. Relevant areas include human ecology, the perception and representation of the environment, ethno-ecological knowledge, the human dimension of biodiversity conservation and the ethnography of environmental problems. While the underlying ethos of the series will be anthropological, the approach is interdisciplinary.

Volume 1
The Logic of Environmentalism
Anthropology, Ecology and Postcoloniality
Vassos Argyrou

Volume 2
Conversations on the Beach
Fishermen's Knowledge, Metaphor and Environmental Change in South India
Götz Hoeppe

Volume 3
Green Encounters
Shaping and Contesting Environmentalism in Rural Costa Rica
Luis A. Vivanco

Volume 4
Local Science vs Global Science
Approaches to Indigenous Knowledge in International Development
Edited by Paul Sillitoe

Volume 5
Sustainability and Communities of Place
Edited by Carl A. Maida

Sustainability and Communities of Place

Edited by
Carl A. Maida

Berghahn Books
New York • Oxford

Dedication

To the memory of Stanley Diamond, Roy A. Rappaport, and Eric R. Wolf

First published in 2006 by
Berghahn Books
www.berghahnbooks.com

©2006, 2011 Carl A. Maida
First paperback edition published in 2011

Library of Congress Cataloging-in-Publication Data

Sustainability and communities of place / edited by Carl A. Maida.
   p. cm. -- (Studies in environmental anthropology and ethnobiology)
   Includes bibliographical references and index.
   ISBN 978-1-84545-016-7 (hbk) -- ISBN 978-0-85745-146-0 (pbk)
   1. Sustainable development. 2. Community life. I. Maida, Carl A.

 HC79.E5S86455 2006
 338.9'27--dc22   2006019281

British Library Cataloguing in Publication Data

A catalogue record for this book is available from the British Library

Printed in the United States on acid-free paper

ISBN 978-1-84545-016-7 (hardback)
ISBN 978-0-85745-146-0 (paperback)
ISBN 978-0-85745-284-9 (ebook)

Contents

List of Tables and Figures

Tables

Figures

Acknowledgements

I would like to thank the Secretariat of the XV International Union of Anthropological and Ethnological Sciences Congress, held in Florence, Italy, in July 2003, with the theme of "Humankind/Nature: Past, Present and Future," for support of the session that would bring together the various authors in this volume for two days of presentations, discussions, and conviviality.

I am grateful to Marion Berghahn for her encouragement and support during the entire process of putting this book together. I would like to express my gratitude to Michael Dempsey and Melissa Spinelli, Production Managers at Berghahn Books in New York, and to copyeditor Lori Rider, for expert guidance on editorial matters. Thanks also to Roy Ellen, who provided a critical reading of an earlier version of the manuscript, and to Barbara Yablon Maida for her close structural reading of earlier versions of each of the essays in this volume and for her creative support during the editing of this book.

Introduction

Carl A. Maida

Sustainability, Local Knowledge, and the Bioregion

The concept of sustainability holds that the social, economic, and environmental factors within human communities must be viewed interactively and systematically. The Brundtland Report (World Commission on Environment and Development, 1987) defines sustainable development as meeting the needs of the present without compromising the ability of future generations to meet their own needs. In 1996, an international group of practitioners and researchers met in Bellagio, Italy, to develop new ways to measure and assess progress toward sustainable development. The Bellagio Principles (1997) serve as guidelines for the whole of the assessment process, including the choice and design of indicators, their interpretation, and communication of the results.

Desmond McNeill (2000) identifies a conflict of interpretations over the concept of sustainable development among various "actors" who have a stake in the sustainability debate. These include academics, activists, and bureaucrats, both legal and political, whose divergent perspectives, interests, and assumptions translate into competing claims as to the very definition of the term. McNeill locates the conflicts among these differing disciplines and professional orientations as arising from earlier debates about poverty, development, and environmental protection, and he calls attention to an emerging conflict between "technicists" seeking technical solutions to environmental problems and "humanists" whose critical perspectives move toward political solutions. Blake Ratner (2004a) calls for an integrated approach toward resolving the polarizing tendencies within the debate. Ratner singles out political economists, rational choice theorists, and cultural institutionalists, positing that their key perspectives, namely equity, efficiency, and cultural identity, respectively, are "three legs of a stool" and therefore necessary elements of any socially acceptable response to sustainability.

Although broadly conceived, the pursuit of sustainable development is a local practice because every community has different needs and quality of life concerns. Despite local variation, the participation of ordinary citizens, or "deliberative democracy," remains constant across the sustainable community movement (Hempel 1998; Agyeman & Angus 2003; Ratner 2004b). In rural areas undergoing rapid development and urban areas transformed by planning, renewal, and clearance, new partnerships are forming on behalf of sustainable development. Residents, and state and nongovernmental organization (NGO) experts, are partnering to design indicators and to monitor land, labor, housing, health, and other quality of life concerns. Civic engagement by ordinary residents is essential as local people have practical experience and bring important intuitive insights to the tasks of indicator design and monitoring. Jane Jacobs (1961) argued on behalf of such "self-diversification," or neighborhood transformation that reflects the vitality, mobility, and aesthetic interests of its residents.

An "ideal" sustainability indicator would be an interactive measure of how long and how well a certain feature of the quality of life or the health of a natural system could maintain itself. Such a measure would require a clear definition of the limits or carrying capacity of a system, or region, or community, to continue on without degrading the environment or decreasing biodiversity while sustaining human and biological life at a certain healthy level. True sustainability indicators will have to wait until both social science and ecological science give us better and measurable models of the operation of interacting systems and until economic measurements can be accurately correlated with social and environmental factors.

However, the interactions and shared effects of economic activity, human society, and the environment can be measured, if not precisely at this stage, at least with some clarity about how they in general improve or degrade each other over time (Bossel 1999; Hart 1999; United Nations 2001). This involves devising custom-made measures of a local system and stipulating from a values perspective what are desirable or undesirable effects. This requires a combination of artfully designed measurement devices and human-defined goals (Haughton & Hunter 1994; Roseland 1998; Meter 1999, 2002). It will also require a better grasp of philosophical and ethical issues surrounding sustainable development (Lee, Holland & McNeill, 2000).

Holistic planning theory has begun to recognize the connection between ecological principles, such as conservation, biodiversity, and restoration, and social scientific principles, such as human scale, social equity, and human development (Archibugi 1998; Beatley 1999; Calthorpe & Fulton 2001). Those who adopt these principles—New Urbanists, advocates of "smart growth" and "livable cities," and proponents of sustainable communities—are rediscovering the vision of "the regional city" as the basis of good and equitable planning and design (Goodman & Goodman 1960; Luccarelli 1995; Duany, Plater-Zyberk & Speck 2000).

The idea of sustainability returns to the notion of the region as the ground of both reason and democracy. Regional culture provides not only a diversity of practices for its citizens to experience but also local perspectives to shape their personal identities. These experiences engender a collective identity and ecology of public symbols that help a community to define place-centered ethical and aesthetic norms. Once internalized, these norms provide individuals with the coherence that fosters more immediate and spontaneous forms of interaction.

This lay form of civic community relies upon a distinct way of knowing that differs from yet is supportive of the universalistic assumptions of the Enlightenment. Clifford Geertz (1983) refers to this way of knowing as "local knowledge" and contends that this form of "fugitive truth" can still be derived from contemporary practices, such as politics, law, ethnography, and poetry, as all are "crafts of place" and "are alike absorbed with the artisan task of seeing broad principles in parochial facts" (Geertz, 1983:167). James Scott (1976), who, like Geertz, has studied Southeast Asian agrarian societies, observed a form of local knowledge in peasant economic, social, and moral arrangements. In the local economies of Burma and Vietnam, Scott found a "subsistence ethic"–or a right to a minimal guaranteed return on the harvest as a hedge against food shortages–embedded in the economic practices and social exchange relations of peasant community life.

Especially valuable in understanding this phenomenon in urban life is the work of social historians and historically informed social scientists interested in the structural aspects of social formations, particularly how their dynamics reflect changes in knowledge and civic culture. Robert Putnam (1993) investigated civic traditions in modern Italian politics and emphasized the primacy of "civic engagement" as a central form of civic virtue. Thomas Bender (1993) studies the extent to which the sense of civic obligation depends upon the character of "urban knowledges." Both seek to document the ways in which "civic community" is represented in what Pierre Bourdieu (1977) would call the habitus of modern citizens, namely the bundle of conventions and habitual ways and perceptions that order lives in particular times and places. Such lay initiatives represent what is perhaps the most fundamental and least appreciated form of civic engagement: the direct and highly active participation of citizens in public affairs by means of NGOs.

An area where considerable attention has been paid to local knowledge is that of bioregionalism, a concept based on ecological principles and traditions of vernacular culture. Bioregionalists envision a more equitable relationship between human and natural systems through reorganizing society around common ecosystems or bioregions and upon sustainable principles. The literature on common pool resources and common property has focused on environmental degradation and resource depletion, and scholars of the commons have offered community-based conservation as a corrective (Agrawal 2003). Some have called for a "recovery of the commons" as a means of regaining local community through peoples' direct involvement in the web of the natural resources (Snyder 1990). This would come about through a revitalized sense of citizenship based upon shared governance around food, water, soil, and energy resources (Orr 1994). Some have even suggested that a "common covenant" could result if people allied themselves by virtue of a common watershed (Jackson 1987). Such a place-focused politics would become viable if local communities were rebuilt upon ecological principles rather than upon political or economic centralization. These "communities of place" would be complex and human-scaled—ones that forge connections between people, foster a sense of well-being, and ensure resilience in crises (Calthorpe & Fulton 2001).

Adaptive Learning and Human Survival

Sustainable development cannot be understood apart from a community, its *ethos,* and its ways of life. Cultural processes, such as norms, values, and expectations, operate as precedents to guide human adaptation, as in the case of a community facing development choices. Culture forms part of a milieu, or adaptive nexus, in which humans learn to cope by taking these precedents into account (Bennett 1996). It acts as a collective memory for human groups to store and retrieve knowledge on which to model future events (Adams 1988). Adaptive behavior, or strategic coping, requires the anticipation of outcomes, using foresight and intentionality as cognitive potentials. In this sense, culture is a form of anticipatory behavior specific to humans, for much of our time is spent reorganizing the world to resist randomness or entropy (Bennett 1996). Humans are goal-directed and self-organizing systems that adapt using new and old information to anticipate outcomes (Moran & Gillett-Netting 2000). The capacity for *internal* adaptation is seen in coping, a short-term process of stress reduction through which individual organisms respond to fluctuations in the environment (Alland 1970).

We also have the capacity to relate with the environment through image-making (Arnheim 1969; Langer 1972). This form of *external* adaptation rests on the ability to symbolically represent the physical world as a cognitive map (Bennett 1996). A cultural worldview reflects the shared cognitive categories of people that experience, and work within, a *local* set of spatial and temporal arrangements. Culture unifies the cognitive maps of different individuals within a locale by imposing "consistency among meanings," as a paradigm for working with the energetic world. By unifying the mental and the energetic in a symbolic system, a local society provides the means to reproduce its own self-organization (Adams 1988).

This form of social reproduction is frequently reinforced through ritual. Anthropology, ethology, and neuroscience view ritualization as adaptive behavior in its ability to encode cultural knowledge (Laughlin, McManus & d'Aquili 1993; Rappaport 1999; Watanabe & Smuts 1999). Ritual symbols, according to Victor Turner (1969), prompt social action because their referents call up polarities between physiological phenomena and normative values, such as reciprocity, respect, generosity, and kindness. Ritual—as a performance—uses these multiple sensory domains in order to provoke an exchange between the physiological and cultural poles, and to reinforce social trust and perhaps a sense of *communitas* among participants.

Living systems theory posits an articulation between humans and the global environment (Miller 1978). This model assumes that human, technical and social spheres are mediated by individual and collective behavior, which is influenced at the biosocial level by physiological and metabolic processes. A living system depends upon subsystem components for its survival, that is, information channels, memory, decision centers, motor outputs, and reproductive elements. All living systems, from cells to the earth, rely on matter, energy, and information flows.

Anthropological holism and living systems theory regard purposive behavior, or culture, as an open system interacting with the environment through positive feedback. The holistic approach views culture as developing in relation to the environmental niche, and social practices as adaptive responses to particular ecological pressures. Historical ecology regards landscape as a manifestation of this dialectical relationship between human action and natural systems over time (Crumley 1994). Living systems theory similarly uses cognitive potentials to explain adaptation in crisis. The systems view holds that organizations and communities, like individuals, will need to draw upon competencies derived from adapting to past crises as sources of feedback. Moreover, how each system experiences a critical event and the emergent pattern of responses will influence the direction of change through later stages of its development.

Using this framework, one may characterize three main evolutionary transitions in human history. Agriculture marked the achievement of long-range predictive control over the food supply through intensive land use techniques. The Industrial Revolution liberated people from this direct symbiotic relationship to the land but severed their local dependence upon the land itself. The second industrial revolution constituted the emergence of human systems capable of large-scale intervention into natural systems, exemplified by human population growth and greater control over the earth's resources. This includes subterranean exploitation of energy and metals; vertical expansion into the atmosphere and oceans for nitrogen, minerals, and food: and control over areas of the electromagnetic spectrum. This last transition brought with it the power to transform the planet into a global system through environmental-exploitative techniques (Bennett 1976). The processes of control in the global system, that is extraction, production, distribution, transportation, and communication, are regulated by diverse ideologies. Global pressures have required local communities to reinvent the symbolic and organizational elements of their cultures in the face of new technologies and ways of life.

At a different scale, local community, or locality, denotes both a physical space and a distinct sensory order where concentrations of people engage in complex networks of social relations (Leeds 1973). Primary relationships of kinship, friendship, and neighborliness, based upon face-to-face interaction, are the most immediate forms of association. Less personal relationships, based upon transaction, are the secondary modes of activity. A loose social organization derives from this multiplicity of contexts, events, and situations. Localities retain their flexible and somewhat amorphous structure because they can accommodate diverse social relationships within their boundaries and control the outcomes of most external intrusions. As highly organized segments of a population, localities support a social structure where individuals take on multiple roles within many cross-cutting networks. Within this configuration, social resources are viewed as potentials and as rights accrued by virtue of a person's status and role within each network. A locality strives to maintain internal control through everyday routines and rituals, and through the networks that govern interpersonal behavior. Local power resides in the internal control of both human and material resources and tends to limit the encroachment of external institutions, such as state or corporate bureaucracies.

The concept of sustainable development, as framed by Redclift (1987), links the transfer of capital, labor, and natural resources within the global economic system. Through a comparative framework that situates the historical role of the environment within capitalist development, Redclift views resource exploitation and structural underdevelopment in the southern hemisphere as a consequence of environmental change in the industrialized northern hemisphere. With global change, localities throughout the world have undergone ecological crises, such as resource depletion, changes in land use, and biodiversity loss. These conditions are frequently accompanied by anthropogenic hazards, such as climate change, greenhouse warming, and emerging epidemic diseases, as well as chaotic environmental episodes such as drought, flooding, and violent storms. Despite efforts to maintain internal control of their economies, many localities become enmeshed in global markets and, as a result, experience increased pressures to change their styles of work and land tenure practices, and to specialize in order to remain competitive. Local communities not only become dependent upon external market forces but are also bound by the policies of development programs designed to introduce technological change. In the past, localities would call upon culture to guide decisions about resource use, as in the case of sustained yield resource management. Locally determined strategies were directed to sustaining an internal equilibrium and were not motivated by demands from outside the local system (Bennett 1976).

Delocalization results when people become less affected with local concerns, especially in decisions about the management of common resources, and in their stance towards their neighbors who have been marginalized by consequences of global change (National Science Foundation 1995). Through its encounter with these displacements, the new ecological anthropology has come to view the community as embedded within larger systems at the regional, national, and international levels, and to study the impact of a multitiered and globalizing world on the locality (Marcus 1995; Gupta & Ferguson 1997; Kottak 1999; Burawoy 2000). This new paradigm recognizes the importance of cultural mediations in ecological processes at a time when local ethnoecologies are being transformed by development, biodiversity conservation, environmentalism, and the influence of NGOs (Brosius 1999; Escobar 1999). Within political ecology, environmental justice research has addressed the ways poor communities organize to confront disproportionate, high, and adverse environmental exposure (Pastor 2001, 2002; Harper & Rajan 2002).

The global economy has led to the transformation of cities, such as New York, London, Tokyo, Sydney, Toronto, Miami, and Los Angeles, into "transnational market spaces" more oriented to world markets than to their national economies (Sassen 1994). Global cities are strategic places in the world economy where the centralized control and management operations required to direct a geographically dispersed array of economic activities are located. As the hubs of global financial markets, these cities are places where there is considerable foreign direct investment and where the broader social structure has grown more international. Their workforces deliver highly specialized services, including finance, telecommunications, and advertising, to diverse linguistic and cultural communities worldwide.

The emergence of globally oriented service industries within these cities, together with the decline of mass production, has created new inequalities and economic polarization. There is a growing earnings disparity between those within the city linked to the international economy and those who remain marginal to it. There is also a disparity in consumption patterns between those employed in the major growth sectors that have high-paying jobs and the low-wage workers employed by small, low-cost service operations. Economic globalization has contributed to a "new geography of centrality and marginality" that elevates certain localities as central to the international economy, while rendering others marginal to the production and distribution of global capital. The turn from the local toward the global resulted in the population movements and dislocations that characterize a crisis, referred to as "late capitalism" or "post-Fordism," first in the developed northern hemisphere, and more recently in the rapidly developing southern hemisphere (Harvey 1989).

Networks and Reorientation

The dislocations caused by the extreme situations of our time have spawned a new pattern of strategic coping, especially when established mechanisms cannot respond adequately to the community's needs. Consequently, various mediating structures–social networks, mutual aid groups, cooperatives, and associations based on ethnic, community, and voluntary ties–move into the vacuum to act as pathways through crisis. These informal social resources promote strategies for survival amid the fragmentation that accompanies dislocation (Fischer 1982). Through them, individuals become cognizant of alternative forms of problem solving, help seeking, and negotiating after crises. Those who embrace these resources gain support and mutual aid, but they also realize an expressive dimension through their participation. The forms of association mobilized during a crisis affect survivors' lives since their sense of belonging reduces the isolation that results from dislocation, and the emergent ties help survivors restore their psychological and social equilibrium. Through facing a crisis and coping with peers in voluntary efforts, survivors learn pragmatic strategies of self-construction. These social milieus cultivate a collective strength and a personal identity, one capable of surviving the multiple crises of late modernity. This form of engagement, called "life politics" by Giddens (1991), concerns issues, such as environmental risks, nuclear power, food security, and reproductive technologies, where self-identity is influenced by globalizing processes.

Networks of civil engagement–mutual aid organizations and other small-scale voluntary associations–were essential to community life in the face of myriad dislocations that marked the onset of modernity in the United States, Western Europe, and elsewhere. Robert Putnam (1993) and his associates have demonstrated, through extensive study of the role of civic traditions in the development of contemporary Italian regional governments, that denser mutual support networks in a community ensure a greater likelihood of civic engagement. Since the medieval period, civic legacies in the towns of northern and central Italy were built upon institutions that supported social solidarity, such as voluntary associations and mutual aid societies. These networks of civic engagement were crucial to the management of collective life in Italian communal republics.

The roots of civic community were thus embedded in a pattern of associational life that traced its ancestry to earlier periods of civic inventiveness. Nineteenth-century Italy saw such a situation as local communities, governed for centuries by civic republicanism, were obliged to develop new forms of collective action for mutual benefit to confront the risks of a rapidly changing social order. In facing the dislocations associated with modernity, localities throughout Italy relied on their civic traditions to guide them in forging a new sense of civic commitment. The emergent "modern" form of civic community, built upon cooperative organizations, cultural associations, and other vehicles for civic mobilization created amid the turmoil of nineteenth-century life, was largely responsible for the success of the regional governments established in northern and central Italy during the 1970s. As they did when their cultural fabrics were rapidly transformed by technological and social advances associated with nation-building and industrialization a century ago, contemporary Italian localities called upon their civic traditions to direct them in the task of reshaping civic culture in the face of regionalization. In these regional governance efforts, cities and their surrounding rural areas framed joint strategies that both stabilized urban fresh food supplies and created clusters of rural industries that extended from the factory to cottage levels. Similar strategies have been adopted in France and Holland, as well as by the European Union.

By comparison, in the United States, massive transformations of civic life during the 1960s and 1970s have brought about considerably different outcomes. For one thing, there has been a tendency to polarize urban areas from rural areas, strengthening transnational market spaces at the expense of sustainable communities and local food networks. Further, the American civic universe has since been characterized by national advocacy organizations that are professionally dominated and far less dependent upon voluntary participation than were the local membership organizations of previous eras (Skocpol 2003). As a result, the public sphere where people actively engage in politics and policy-making has changed (Calhoun 1992). In the United States and elsewhere in the North, the move away from local associations to global advocacy organizations has had an impact on local communities worldwide. Until recently, these localities have used cultural dynamics similar to the ones Putnam describes to survive the transition to urban, industrial life in the nineteenth and twentieth centuries, and to confront the turbulence of globalization in our own times.

However, there have also been changes in the global "Third Sector" as a result of the transformation of NGOs within national societies of the North (Salamon & Anheier 1997; Anheier & Toepler 1999). In the area of sustainability and natural resource management, significant amounts of international donor monies have been appropriated for interventions on behalf of biodiversity conservation and land stabilization (Brosius, Tsing & Zerner 1998). Consequently, NGO-initiated community-based conservation projects frequently conflate the local knowledge of indigenous resource-owning residents with the well-meaning conservationist initiatives of transnational NGOs (Igoe 2003; Hviding 2003). Through these projects, local-level movements for control over natural resources frequently link up with NGO-inspired transnational advocacy networks. Much of the recent debate, emerging from the convergence of local and global interests through these movements, involves the way each set of actors has come to interpret local knowledge (Dumoulin 2003).

The cases that comprise the book further explore the ways local communities have reinvented themselves using cultural knowledge to blend traditional sentiments with fully modern sensibilities, and to sustain both local and regional networks and the sense of cultural identity amid large-scale dislocations within their own societies and in the international economy. Therefore, following the lead of Eric Wolf (1982) and other anthropologists and social scientists, in general, who have been "rediscovering" history in the social sciences, the authors attempt to place recent examples of social formations related to sustaining natural resources and preserving cultural knowledge into a larger context.

Local and Global Knowledges

The first set of essays focuses on conflicts between local and global knowledges in biomes, such as tropical rainforests and rivers. Claude Raynaut and his coauthors provide a critical analysis of the notion of sustainability, showing that, even when applied to a local place and to issues of protection and preservation of nature and of local cultures, it cannot be established upon clear, ready-made criteria that could guide an attempt to reach a stable, long-lasting situation of natural and social equilibrium. Thomas Thornton looks at problems that come about when indigenous communities develop regional and village-level business corporations to manage land, resources, and income. These local practices will often conflict with state regulations to conserve wild, renewable resources, such as fish, plants, and wildlife, as a legal way to enhance their sustainability. Johanna Gibson calls for an expansion of the concept of "place-centered community" to include collective values, such as shared identity and mutual responsibility in facing the potential loss of connection to the land. In other words, "community" can be understood as including indigenous biological resources and an ethos of preservation of cultural diversity within a globalizing framework of biodiversity. Dario Novellino argues that despite expanded notions of community that include ecological sustainability and protection of biodiversity, there remains a threat to local knowledge and practices because of conflicting priorities. As a result of their efforts to promote local knowledge, both government and NGO conservationists come into conflict with the ways of perceiving and using natural resources by traditional communities. On the other hand, the indigenous people themselves have difficulties in conceiving of a unitary notion of "culture" and "community" that could serve a political function.

Local Practices: Adaptive Strategies and State Responses

This set of essays concerns local practices in communities in transition, focusing on the conflicts between innovative adaptive strategies used by local communities to preserve resources and ways of life, and regulative responses by state agencies. Krista Harper analyzes environmental politics and activism during the post-socialist transformation of Central and Eastern Europe. The spatial dimensions of urban-rural inequalities and marginalization are presented in the context of both a growing environmental awareness and new advocacy organizations among ethnic populations. The ensuing tension between environmental justice and mainstream environmentalism challenges existing notions of sustainability. Deborah Pellow describes how the dual processes of transnational migration and the preservation of sustainable foodways help maintain the informal sector of urban ethnic enclaves. Many ethnic neighborhoods have local markets that sell prepared ethnic foods made by migrants from rural and tribal areas. In the cosmopolitan neighborhoods of Third World cities, one finds a "creolization" of food consumption in that both local and Western cuisines are co-present in these "globalized local communities." State public health agencies have become increasingly concerned with the conditions under which such foods are produced, stored, and distributed. Janet Benson discusses how the evolution of agricultural practices in the 1960s provided a basis for the development of rural "industrialization," namely meatpacking and agribusiness. This trend was coupled with the use of transnational migrant labor and led to municipal growth in farming areas. As a result, state services have increased, since town dwellers have greater needs for public health, education, and social services. Current practices have led to major water conservation issues and threaten agricultural sustainability in the region. Barbara Yablon Maida and Carl Maida examine residents' attitudes toward land use and development and the perceptions of the visual landscape in a rapidly urbanizing agricultural area. After voters passed measures that would ideally control development infringement onto agricultural land and preserve open space, residents and experts designed quality of life indicators to measure features of the economy, social well-being, and environmental health. Civic engagement was crucial to the indicator design process.

Social Capital, Civic Engagement, and Globalization

The third set of essays regards practices that tend to increase social capital in local communities, such as civic engagement and the design of sustainable development indicators. Critical perspectives on cultural practices of consumption and on community self-sufficiency are offered to assist localities in meeting the challenges of globalization. Kenneth Meter describes how residents, technical experts, and professional researchers formed new urban partnerships to address environmental concerns and prevent pollution, and in the process devised new tools that are applicable globally. In an effort to engage residents directly in defining indicators of neighborhood sustainability for their own communities, participants defined linkages among issues that are typically viewed as separate. Karla Caser looks at dynamics in coastal communities as new businesses, primarily ecotourism and commercial travel and tourism, and upwardly mobile residents enter the area. This influx has led to an altered landscape, as well as changes in both class relations and lifestyles that have held the community together for generations. A case study looks at how changes in the physical environment hinder community identity and channel people's use of space, decreasing social capital especially for low-income residents. Bourdieu's praxeology is used to develop an account of the ways by which the built environment objectifies social capital. Such "physical-social capital" symbolically and physically constrains social interaction and engenders shared identity and predictability. The framework is considered especially useful for design professionals, equipping them with critical tools for creating environments responsive to society's contemporary needs while building inclusive communities of place.

Two essays address theoretical issues related to the ongoing debate on globalization and local knowledge. Richard Westra reflects on how the abstract and impersonal effects of globalization on the human "life-world" tend to neglect the human cultural goals of caring, mutual aid, accountability, and shared governance. As a corrective and perhaps reparative strategy, communities of place can attempt to re-embed their local economies in the "life-world," through sustainable practices that increase self-sufficiency and personal autonomy. In this way, local communities can potentially insulate themselves from the ravages of global markets. Snježana Čolić discusses the prospect of sustainability within a globalizing world and how global culture has come to eclipse local knowledge with respect to resource needs and thereby moved localities to embrace more universalist consumption practices, including media and other knowledge commodities of advanced capitalist societies. With these practices come forms of global knowledge that may even include new styles of citizenship and regulations supporting resource conservation, biodiversity, and sustainable development. Moreover, the new media conveying these commodities create a "cultural space of the global" that is relatively void of context; the attendant cognitive dissonance, anxiety, and insecurity will often provoke a sense of cultural disenchantment. To critically analyze the various ways sustainable ideas and practices, in the context of globalization, have brought about dynamic changes in local communities, as the papers in this volume demonstrate, anthropologists will have to approach the study of culture through multiple perspectives, based upon the interests and needs of particular societies, rather than the universalist interests of any single ideological, historical, or methodological tradition.

References

Adams, Richard Newbold. 1988. *The Eighth Day: Social Evolution as the Self-Organization of Society.* Austin: University of Texas Press.

Agrawal, Arun. 2003. "Sustainable Governance of Common-Pool Resources: Context, Methods, Politics." *Annual Review of Anthropology* 32:243-62.

Agyeman, Julian, and Briony Angus. 2003. "The Role of Civic Environmentalism in the Pursuit of Sustainable Communities." *Journal of Environmental Planning and Management* 46, (3): 343-63.

Alland, Alexander. 1970. *Adaptation in Cultural Evolution: An Approach to Medical Anthropology.* New York: Columbia University Press.

Anheier, Helmut K., and Stefan Toepler, eds. 1999. *Private Funds, Public Purpose: Philanthropic Foundations in International Perspective.* New York: Plenum.

Archibugi, Franco. 1998. *The Ecological City and the City Effect: Essays on the Urban Planning Requirements for the Sustainable City.* London: Avebury Publishers.

Arnheim, Rudolf. 1969. *Visual Thinking.* Berkeley: University of California Press.

Beatley, Timothy. 1999. *Ethical Land Use: Principles of Policy and Planning.* Baltimore: The Johns Hopkins University Press.

Bellagio Principles: Guidelines for the Practical Assessment of Progress Towards Sustainable Development. 1997. Winnipeg, Manitoba, Canada: International Institute for Sustainable Development.

Bender, Thomas. 1993. *Intellect and Public Life: Essays on the Social History of Academic Intellectuals in the United States.* Baltimore: The Johns Hopkins University Press.

Bennett, John W. 1976. *The Ecological Transition: Cultural Anthropology and Human Adaptation.* New York: Pergamon.

___. 1996. *Human Ecology as Human Behavior: Essays in Environmental and Development Anthropology.* New Brunswick: Transaction Books.

Benson, Janet E. 2007. "Globalization, Local Practice, and Sustainability in the High Plains Region of the United States." In *Sustainability and Communities of Place,* ed. Carl A. Maida. Oxford and New York: Berghahn Books.

Bossel, Hartmut. 1999. *Indicators for Sustainable Development: Theory, Method, Applications.* Winnipeg, Manitoba, Canada: International Institute for Sustainable Development.

Bourdieu, Pierre. 1977. *Outline of a Theory of Practice.* Translated by Richard Nice. Cambridge: Cambridge University Press.

Brosius, J. Peter. 1999. "Analyses and Interventions: Anthropological Engagements with Environmentalism." *Current Anthropology* 40 (3): 277-309.

Brosius, J. Peter, Anna Lowenhaupt Tsing, and Charles Zerner. 1998. "Representing Communities: Histories and Politics of Community-Based Natural Resource Management." *Society and Natural Resource Management* 11 (2): 157-68.

Burawoy, Michael, et al. 2000. *Global Ethnography: Forces, Connections, and Imaginations in a Postmodern World.* Berkeley: University of California Press.

Calhoun, Craig, ed. 1992. *Habermas and the Public Sphere.* Cambridge, MA: MIT Press.

Calthorpe, Peter, and William Fulton. 2001. *The Regional City: Planning for the End of Sprawl.* Washington, DC: Island Press.

Caser, Karla. 2007. "The Design of the Built Environment and Social Capital: Case Study of a Coastal Town Facing Rapid Changes." In *Sustainability and Communities of Place,* ed. Carl A. Maida. Oxford and New York: Berghahn Books.

Čolić, Snježana. 2007. "The Prospect of Sustainability in the Culture of Capitalism, Global Culture and Globalization: A Diachronic Perspective." In *Sustainability and Communities of Place*, ed. Carl A. Maida. Oxford and New York: Berghahn Books.

Crumley, Carole L. 1994. "Historical Ecology: A Multidimensional Ecological Orientation." In *Historical Ecology: Cultural Knowledge and Changing Landscapes*, Carole L. Crumley, ed., 1-16. Santa Fe, NM: School of American Research Press.

Duany, Andres, Elizabeth Plater-Zyberk, and Jeff Speck. 2000. *Suburban Nation: The Rise of Sprawl and the Decline of the American Dream*. New York: North Point Press.

Dumoulin, David. 2003. "Local Knowledge in the Hands of Transnational NGO Networks: A Mexican Viewpoint." *International Social Science Journal* 55 (178): 593-606.

Escobar, Arturo. 1999. "After Nature: Steps to an Antiessentialist Political Ecology." *Current Anthropology* 40 (1): 1-30.

Figueroa, Mary Elena, D. Lawrence Kincaid, Manju Rani, and Gary Lewis. 2002. *Communication for Social Change: An Integrated Model for Measuring the Process and Its Outcomes*. New York: The Rockefeller Foundation.

Fischer, Claude S. 1982. *To Dwell Among Friends: Personal Networks in Town and City*. Chicago: University of Chicago Press.

Geertz, Clifford. 1983. "Local Knowledge: Fact and Law in Comparative Perspective." In *Local Knowledge*. New York: Basic Books.

Gibson, Johanna. 2007. "Communities Out of Place." In *Sustainability and Communities of Place,* ed. Carl A. Maida. Oxford and New York: Berghahn Books.

Giddens, Anthony. 1991. *Modernity and Self-Identity: Self and Society in the Late Modern Age*. Stanford, CA: Stanford University Press.

Goodman, Paul. and Percival Goodman. 1960. *Communitas: Means of Livelihood and Ways of Life*. New York: Random House.

Gupta, Akhil, and James Ferguson, eds. 1997. *Anthropological Locations: Boundaries and Grounds of a Field Science*. Berkeley: University of California Press.

Harper, Krista. 2007. "Does Everyone Suffer Alike? Race, Class, and Place in Hungarian Environmentalism." In *Sustainability and Communities of Place,* ed. Carl A. Maida. Oxford and New York: Berghahn Books.

Harper, Krista, and S. Ravi Rajan. 2002. *International Environmental Justice: Building the Natural Assets of the World's Poor*. Political Economy Research Institute. International Natural Assets Conference Paper Series 12. Amherst, MA: University of Massachusetts.

Hart, Maureen. 1999. *Guide to Sustainable Community Indicators*, 2nd ed. North Andover, MA: Hart Environmental Data.

Harvey, David. 1989. *The Condition of Postmodernity: An Enquiry into the Origins of Cultural Change*. Oxford: Blackwell.

Haughton, Graham, and Colin Hunter. 1994. *Sustainable Cities*. Regional Policy and Development Series, No. 7. London: Taylor and Francis.

Hempel, Lamont C. 1998. *Sustainable Communities: From Vision to Action*. Claremont, CA: Claremont Graduate University.

Hviding, Edvard. 2003. "Contested Rainforests, NGOs, and Projects of Desire in Solomon Islands." *International Social Science Journal* 55 (178): 539-53.

Igoe, Jim. 2003. "Scaling Up Civil Society: Donor Money, NGOs and the Pastoralist Land Movement in Tanzania." *Development and Change* 34 (5): 863-85.

International Food Policy Research Institute. 2002. *Sustainable Food Security for All by 2020—Proceedings from an International Conference, September 4-6, 2001–Bonn, Germany.* Washington, DC.

Jackson, Wes. 1987. "Toward a Common Covenant." In *Altars of Unhewn Stone: Science and the Earth.* San Francisco: North Point Press.

Jacobs, Jane. 1961. *The Death and Life of Great American Cities.* New York: Random House.

Kottak, Conrad P. 1999. "The New Ecological Anthropology." *American Anthropologist* 101 (1): 22-35.

Langer, Suzanne K. 1972. *Mind: An Essay on Human Feeling,* vol 2. Baltimore: The Johns Hopkins University Press.

Laughlin, Charles D., John McManus, and Eugene G. d'Aquili. 1993. *Brain, Symbol, and Experience: Towards a Neurophenomenology of Human Consciousness.* New York: Columbia University Press.

Lee, Keekok, A.J. Holland, and Desmond McNeill, ed. 2000. *Global Sustainable Development in the Twenty-first Century.* Edinburgh: Edinburgh University Press.

Leeds, Anthony. 1973. "Locality Power in Relation to Supralocal Power Institutions." In *Urban Anthropology: Cross-Cultural Studies of Urbanization,* ed. Aidan Southall. New York: Oxford University Press.

Luccarelli, Mark. 1995. *Lewis Mumford and the Ecological Region: The Politics of Planning.* New York: The Guildford Press.

Maclaren, Virginia W. 1996. *Developing Indicators of Urban Sustainability: A Focus on the Canadian Experience.* Toronto: ICURR Press.

Maida, Barbara Yablon, and Carl A. Maida. 2007. "Quality of Life, Sustainability, and Urbanization of the Oxnard Plain, California." In *Sustainability and Communities of Place,* ed. Carl A. Maida. Oxford and New York: Berghahn Books.

Marcus, George. 1995. "Ethnography in/of the World System: The Emergence of Multi-Sited Ethnography." *Annual Review of Anthropology* 24: 95-117.

McNeill, Desmond. 2000. "The Concept of Sustainable Development." In *Global Sustainable Development in the Twenty-first Century,* ed. Keekok Lee, A. J. Holland, and Desmond McNeill. Edinburgh: Edinburgh University Press.

Meter, Ken. 1999. *Neighborhood Sustainability Indicators Guidebook.* Minneapolis, MN: Crossroads Resource Center.

___. 2002. *Food with the Farmer's Face on It: Emerging Community-Based Food Systems.* Battle Creek, MI: W. K. Kellogg Foundation.

___. " 2007. Linked Indicators of Sustainability Build Bridges of Trust." In *Sustainability and Communities of Place,* ed. Carl A. Maida. Oxford and New York: Berghahn Books.

Miller, James Grier. 1978. *Living Systems.* New York: McGraw-Hill.

Moran, Emilio F. and Ronda Gillett-Netting, eds. 2000. *Human Adaptability: An Introduction to Ecological Anthropology,* 2nd ed.. Boulder, CO: Westview Press.

National Science Foundation. 1995. *Cultural Anthropology, Global Change and the Environment: Anthropology's Role in the NSF Initiative on Human Dimensions of Global Change.* Report of a Workshop on Human Dimensions of Global Change. Washington, DC, June 27-28.

Novellino, Dario. 2007. "'Talking about Kultura and Signing Contracts': The Bureaucratization of the Environment on Palawan Island (the Philippines)." In *Sustainability and Communities of Place,* ed. Carl A. Maida. Oxford and New York: Berghahn Books.

Orr, David W. 1994. *Earth in Mind: On Education, Environment and the Human Prospect.* Washington, DC: Island Press.

Pastor, Manuel. 2001. *Building Social Capital to Protect Natural Capital: The Quest for Environmental Justice.* Political Economy Research Institute. Working Papers Series Number 11. Amherst, MA: University of Massachusetts.

____. 2002. *Environmental Justice: Reflections from the United States.* Political Economy Research Institute. International Natural Assets Conference Paper Series 1. Amherst, MA: University of Massachusetts.

Pellow, Deborah. 2007. "Attachment Sustains: The Glue of Prepared Food." In *Sustainability and Communities of Place,* ed. Carl A. Maida. Oxford and New York: Berghahn Books.

Putnam, Robert D. 1993. *Making Democracy Work: Civic Traditions in Modern Italy.* Princeton, NJ: Princeton University Press.

Rappaport, Roy. 1999. *Religion and Ritual in the Making of Humanity.* Cambridge: Cambridge University Press.

Ratner, Blake D. 2004a. "Equity, Efficiency, and Identity: Grounding the Debate Over Population and Sustainability." *Population Research and Policy Review,* 23 (1): 55-71.

____. 2004b. "Sustainability as a Dialogue of Values: Challenges to the Sociology of Development." *Sociological Inquiry* 74 (1): 50-69.

Raynaut, Claude, Magda Zanoni, Angela Ferreira, and Paulo Lana. 2007. "Sustainability: Where, When, for Whom? Past, Present, and Future of a Local Rural Population in a Protected Natural Area (Guaraqueçaba, Brazil)." In *Sustainability and Communities of Place,* ed. Carl A. Maida. Oxford and New York: Berghahn Books.

Redclift, Michael. 1987. *Sustainable Development: Exploring the Contradictions.* London and New York: Routledge.

Roseland, Mark. 1998. *Toward Sustainable Communities: Resources for Citizens and Their Governments.* Gabriola Island, British Columbia, Canada: New Society Publishers.

Salamon, Lester M., and Helmut K. Anheier, eds. 1997. *The Nonprofit Sector in the Developing World: A Comparative Analysis.* Manchester: Manchester University Press.

Sassen, Saskia. 1994. *Cities in a World Economy.* Thousand Oaks, CA: Pine Forge Press.

Scott, James C. 1976. *The Moral Economy of the Peasant: Rebellion and Subsistence in Southeast Asia.* New Haven: Yale University Press.

Skocpol, Theda. 2003. *Diminished Democracy: From Membership to Management in American Civil Life.* Norman: University of Oklahoma Press.

Snyder, Gary. 1990. "The Place, the Region, and the Commons." In *The Practice of the Wild.* San Francisco: North Point Press.

Thornton, Thomas F. 2007. "Alaska Native Corporations and Subsistence: Paradoxical Forces in the Making of Sustainable Communities." In *Sustainability and Communities of Place,* ed. Carl A. Maida. Oxford and New York: Berghahn Books.

Turner, Victor. 1969. *The Ritual Process.* Chicago: Aldine.

United Nations Division for Sustainable Development. 2001. *Indicators of Sustainable Development: Guidelines and Methodologies.* New York.

United Nations Division for Sustainable Development. 2003. *The Road from Johannesburg: World Summit on Sustainable Development–What Was Achieved and the Way Forward.* New York.

Watanabe, John M., and Barbara B. Smuts. 1999. "Explaining Religion without Explaining It Away: Trust, Truth, and the Evolution of Cooperation in Roy A. Rappaport's 'The Obvious Aspects of Ritual.'" *American Anthropologist,* 101 (1): 98-112.

Westra, Richard. 2007. "Sociomaterial Communication, Community, and Ecosustainability in the Global Era." In *Sustainability and Communities of Place,* ed. Carl A. Maida. Oxford and New York: Berghahn Books.

Wolf, Eric R. 1982. *Europe and the People without History.* Berkeley: University of California Press.

World Commission on Environment and Development. 1987. *Our Common Future* (The Brundtland Report). Oxford: Oxford University Press.

World Conservation Union, United Nations Environment Program, World Wide Fund for Nature. 1991. *Caring for the Earth: A Strategy for Sustainable Living.* Gland, Switzerland.

Part One

LOCAL AND GLOBAL KNOWLEDGES

Sustainability

*Where, When, for Whom? Past, Present,
and Future of a Local Rural Population
in a Protected Natural Area
(Guaraqueçaba, Brazil)*

Claude Raynaut, Magda Zanoni,
Angela Ferreira and Paulo Lana

A rethinking of development models has occurred during the latter half of the twentieth century. For the most part, this rethinking has been critical of purely economic models that only emphasize growth and do not take into account the needs and aspirations of the affected population and risks to the natural environment. Perroux's works (1961), those of the "Club of Rome" (Beckerman 1972; ONU/EPHE 1972) and, lately, Sachs' works (1990) bear witness to this shift in thinking. In the 1990s, the concept of sustainable development came to the forefront. The World Conservation Union (IUCN) first used this notion in its global strategy of conservation (Jacobs and Munro 1987). It was then taken up again and developed by the Brundtland Report (WCED 1987). Despite the successes of the 1992 World Conference on Environment, sustainable development still remains rather blurred and is beset by many differences of interpretation. It is marked by contradictions (Pezzey 1989). No fewer than sixty different definitions have been noted by Latouche (1995). From a development perspective, it integrates a number of dimensions that were not initially included: the acknowledgment that available resources are not infinite, the recognition of the value of biodiversity, the demand for social equity, and a concern for long-term and intergenerational solidarity. It is easy to reach agreement on some statements of principle, but consensus can hardly be found when looking for concrete definitions, identification of priority actions, and the formulation of criteria necessary for evaluating development policy. Vagueness remains, particularly with regard to scientific concepts that could support the notion of sustainability. Among others, different French authors (Dubois and Mahieu 2002; Elame 2001; Jollivet 2001, Martin 2002) have already started the necessary work of critical reflection.

Without going into the details of this discussion, some points should be emphasized that may prove useful when analyzing the case below. We start from a basic assumption: despite the ambiguities of the notion of sustainability and the diversity of its interpretations, almost all definitions are based on the central notion of equilibrium. Attaining "sustainability" often entails succeeding in maintaining stability among forces liable to lead to degradation (van der Leeuw 1998). The criteria characterizing this equilibrium point could change according to the various definitions held. Thus, depending on which point of view is adopted to evaluate the sustainability of a development program or an environmental project, stress could be placed on its economic viability, on the preservation of fragile ecological balances, or else on the necessity to ensure suitable, healthy, and equitable access to natural or human-produced resources for different categories of producers and consumers both now and in the future. In all cases, the goal is the same: finding the means to reach a good balance with regard to the criteria that are taken into account, and acting in order to maintain the "ideal" point so reached. Among such a diversity of approaches, contradictions may appear between the various definitions of sustainability that are adopted, as well as between the principles that can be put forward when it comes to implementing concrete policies.

Contradictions among the various conceptions of balance are particularly obvious with regard to nature protection policies. Indeed, the idea that underlies, more or less exclusively or radically, the project of protecting nature is that each living being has an intrinsic value, and life in all its forms must be preserved for its own sake regardless of any other consideration (Rolston 1994). Such a position raises numerous critiques from those who start from an opposite philosophical and ethical position and who consider that only human beings can be the source and the reference when assigning value to any other beings or objects that are part of the universe. From a practical perspective, there are also those who criticize the negative consequences of preservation policies on areas where the population consists of small familial producers, such as farmers, fishers, and hunter-gatherers, with little capacity to resist policies imposed from more powerful agents (Little 1999; Rossi 2001). What justifies placing more value on an ecosystem than upon the human beings who live in it?

If one agrees that not only ecosystems but also social systems have to be preserved, what does that mean and imply? There is general consensus about the following statement: preservation of human groups cannot be limited to the physical survival of their members but must include the defense of their collective culture, values, and identity. Debates arose over the notion of "tradition" (and "traditional populations") as one of the forms under which the notion of sustainability can be translated in the social field. These debates led to the emergence of the notion of "ethnoconservation" and the protection of "traditional" cultures. The defense of cultural diversity became an objective equivalent to those of nature protection and the preservation of biological diversity (Diegues 2000). This vision has been progressively integrated by some of the conservationist movements, through a valorization of the traditional populations and their traditional knowledge, seen then as potential contributors in the fight to preserve nature. An ideal model of sustainability thus turned upon the notion of the harmony and equilibrium between a traditional population and the natural habitat in which it has been living for ages.

When it comes to implementing concrete policies at a local level, contradictions often occur between the objectives of nature conservation and those of preserving the physical and social integrity of the human groups living within the concerned territories. These contradictions are particularly revealing of the ambiguities that are hidden beneath the apparent consensus reached around the notion of "sustainable development." Even if this issue constitutes only one of the possible angles for a critical analysis of the notion, we will start from it in this paper in order to develop our thinking. It is justified because, since its origins, the idea of sustainable development has been closely associated with the idea of conservation as was recommended by the World Conservation Union in its "global conservation strategy" (Diegues, 1996; World Conservation Union, 1980). It is also inspired by the nature of the case that will be analyzed subsequently: the story of a local policy for environment preservation in Brazil that also claims, at least in some of its stages, to protect the culture of the social groups who have been living there for a long time. Let us take the idea that integrity of nature and sustainability of the relationship that people have with it in a given space require preserving both the stability of natural systems and that of local societies in their tradition. Our point here is that conflating "sustainability" with "stability" lacks a scientific basis, both from the perspective of the natural sciences and from that of the social sciences.

The Variability of Natural Systems

The scientific analysis of natural systems is now diverging more and more from the idea of static balance in order to integrate notions of transformation dynamics, including variability, uncertainty, and reversibility. The concept of resilience, coming from modern physics, has been applied to the analysis of natural systems for some time. Like sustainability, it also is bedeviled by numerous definitions (DeAngelis & Waterhouse 1987; Holling 1976; Ludwig, Walker & Holling 1997; Pimm 1991). Here we will apply the one given by van der Leeuw (1998: 56): "the ability of a system to keep its structure in the face of disturbances, to absorb and adapt to change." This concept rejects a static interpretation of sustainability. The stronger the ecosystem resilience, the more it is able to assimilate changes in the face of external destabilizing forces or internal conflicts and the less it is threatened by extinction. When a disturbance is strong enough to push a biological system beyond its threshold of resilience, it is liable to disappear and to be replaced by a new system with a different composition and structure but which will occupy the same space.

Reproduction and Change: Complexity and Flexibility in the Social Field

Concepts of stability and equilibrium are as pertinent for the analysis of social systems as they are for natural systems. This is largely admitted in cases of great industrial and market-oriented societies whose conflicting dynamics are openly displayed in a national and supranational framework. It is also the case for small, so-called traditional, rural societies, although they have been sometimes spoken of as societies "without history" or as "cold" societies. Balandier (1989) has discussed this matter from a theoretical point of view, showing that disorder, innovation, and change are at the core of every social system.

All human societies are historical. They have undergone changes that were more or less profound or fast, depending on their own characteristics and on the events that have marked their specific history. "Traditional" societies have never been isolated from this dynamic, even if, more recently, colonialism and, later, cultural and economic globalization have strongly accelerated the rhythm of these changes. Tradition has always been the result of a continuous reconstruction (Amselle 1990).

The Relativity of the Notion of Sustainable Development: Diversity of Scales and Actors' Points of View

As soon as this dynamic and historical vision of social and natural systems is adopted, a single preestablished definition of what constitutes "sustainability" loses its meaning. Indeed, the judgment about the sustainable or nonsustainable nature of a particular situation and of a given process of change cannot be expressed in the absolute, from strictly objective and scientific data. It has necessarily to be seen in relation to social, political, and ethical considerations. It must integrate negotiations and arbitrations that take place around two criteria: the scales of space and time and the diversity of the social agents.

The Scales of Space and Time

The nature of the points of view that can be adopted, and actions that can be undertaken in the name of sustainability, are closely dependent on the way of locating oneself according to spatial and temporal frames of reference, which can be schematically represented by the intersection of two axes. The first is a temporal axis, which distinguishes the "present" from the "future." The second is a spatial axis that distinguishes between the "here" (the locality) and all that which is exterior to it, namely, the "elsewhere," which may include different spatial scales according to each particular case.

Therefore, the resolution of a local and current crisis can occur through the displacement of problems toward the exterior or toward the future, for example, the processing of waste. Conversely, broad measures designed to benefit an entire nation or future generations can result in local problems that seriously affect the existence of a particular population, or category of population, or its relationships to the environment, as often happens when regulations for nature protection are undertaken. "Now and here" sustainability is not always reconcilable with the one we might wish for: "elsewhere and in the future." In this case, input into the decision-making process cannot only come from scientific criteria. It has necessarily to be the result of negotiations among diverse points of views of social agents.

The Diversity of the Social Agents

The choices concerning the relevance of changes in the organization of a social group, particularly with regard to its relations with natural systems, are dependent on social agents' points of view. Therefore, the question of sustainability cannot be dealt with as if it had general pertinence. It only has meaning if we ask at the same time, "sustainability for whom?" Thus, a development that allows a given social category to perpetuate or reinforce itself can threaten the survival of another social category. For example, the consumers' desire for more healthy products can menace the economic viability of highly technically intensified farming systems. As soon as the dimension of social reproduction is integrated into the definition of "sustainability," it cannot be isolated from the question of power, that is, the forms of political, social, and economic inequality.

In order to make progress through such a theoretical reflection and to clarify a certain number of questions that have arisen, the link must be made to detailed analysis of specific situations. This is the subject of the research program targeted at an area of conservation in the south of Brazil: the *Área de Proteção Ambiental,* or Area of Environmental Protection (APA) of Guaraqueçaba, in the State of Paraná. Starting from an historical study of the evolution of the local society in its relationship to the natural environment during the twentieth century, we examine the conditions in which the environmental preservation policy was conceived and locally implemented. We show the different points of view and the stakes involved in the conflict, and we inquire into the various agents' representations and practices with respect to the notion of sustainable development and its application. The value of this example comes from the fact that it is a particularly illustrative local case study, both revealing in terms of analysis and capable of providing elements to inform reflection on a wider scale.

The Relationship of Society to Nature in Guaraqueçaba: A Century of Turbulent History

The northern coastal area of the Brazilian State of Paraná is characterized by unique economic, environmental, geographic, and historical features. Separated from the high inland plains of the state by a coastal mountain range (the Serra do Mar), it enjoys a humid subtropical climate (see Figure 1). The most significant remnants of the wet ombrophilous Atlantic forest (Mata Atlântica) can still be found here today. The forest extended over a 3,500-kilometer strip of coastline and covered one million square kilometers at the time of the arrival of the Portuguese. Today, it has practically disappeared (Dean 1995).

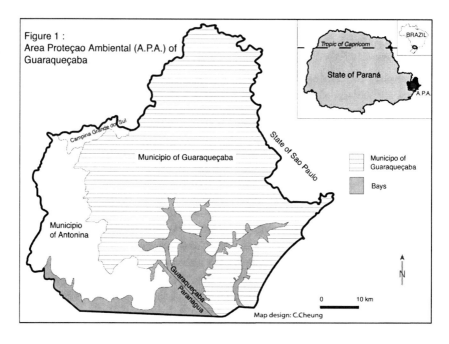

Figure 1. Area Proteçao Ambental (APA) of Guaraqueçaba

Unlike most regions in the south of the country, and unlike the rest of the Paraná region in particular, which were populated by colonization movements during the second half of the nineteenth century, a large part of the agrarian society of this coastal area is rooted in a peasantry whose origins date to the earliest times of colonization, namely the early seventeenth century. Moreover, to a large extent, it has remained apart from the "Green Revolution" that deeply marked the agriculture of the interior during the past half-century. In spite of this relative isolation, major changes during the past century have affected the relationship between all sectors of the local society and the natural environment they exploit. Let us try to recall the major stages of this history.[1]

The Beginnings of a Commercial Agriculture

Occupied and exploited since the seventeenth century by the Portuguese, the northern coast of Paraná was initially dominated by gold prospecting in the streams descending the Serra do Mar. Later, the collection of forest products and agriculture (on both large and small holdings) depended on the availability of servile labor. The abolition of slavery in Brazil in 1888 radically changed this agriculture. The large estates could not survive, and part of the land that they occupied reverted to the statute of "vacant lands and without a Master." Between the first and third decades of the twentieth century, the region experienced a new period of prosperity due to the rise of banana cultivation, whose production was intended for Argentina and Uruguay. These countries owed their prosperity to meat and cereal exports to Europe. This period was characterized by three types of phenomena: (1) a strong influx of immigrants; (2) an increase of the cleavages between small farmers and a minority of well-established landowners; and (3) a widespread clearing of the forests in the alluvial valleys, secondary valleys, and slopes.

The Crisis of the Local Agriculture

After World War II, the northern coast experienced increasing competition from other areas of production in Brazil. During the mid-1930s, exports to countries of the Rio Plata definitively ceased. The area entered a deep economic depression that can be broadly summarized as follows: (1) population migration from the interior of the area toward the coast, which offered a broader range of activities (fishing and urban labor opportunities); (2) increased marginalization of small family farmers; (3) decline in the demand for land for agriculture, but intensification in the collection of palm hearts (*Euterpe edulis*) from the Mata Atlântica.

The 1950s marked a turning point, as economic operators from outside the coastal area began to express interest in some of the local resources–in particular, raw materials for charcoal production or pulp for paper, fine wood from the Mata Atlântica forests, and plots of *Euterpe edulis*. These newcomers seized vacant lands by legal means or with the complicity of the legal and bureaucratic apparatus. In appropriating by legal artifice the land of assignees that were no longer present, and by evicting (often by force) small farmers with a precarious hold on the land, some newcomers succeeded in building "latifundia," consisting of thousands of hectares of well-situated lands, particularly in the valleys and on the lower slopes of the mountains.

By the 1970s, this dynamic had accelerated through federal programs with tax incentives for forest development and agro-industrial activities. The area's attraction was further reinforced by the construction of a drivable road, which opened up the northern coast and connected it to major economic and urban centers. Tension over the land increased as large-scale forest development for timber, charcoal, and palm tree harvesting, as well as mining activities, rapidly degraded the environment. At the end of this period of transformation, the *município* of Guaraqueçaba experienced an exceptionally high level of land concentration, even by Brazilian standards. In 1985, ten latifundia alone proclaimed rights to more than 87 percent of the exploited lands in the commune; their validity could be contested but they were exerted in practice. Such a situation inevitably gave rise to serious social strain. Violent conflicts multiplied between large landowners and small farmers. In addition to the social impact, which was dramatic for small family producers, this evolution, strongly marked by the economic logic of new actors, had serious consequences for the natural environment, particularly the forest, as a result of extensive clearings and overexploitation of palm hearts.

The Rise of the Environmental Agenda

At the beginning of the 1980s, Brazilian political life witnessed a movement for profound change. The transition between a military dictatorship and a civil power was near. The economic model, which had inspired the "Brazilian miracle," was subjected to sharp criticism from actors representing a range of social movements. Many of these movements placed priority on arguments for environmental defense and for the quality of life in order to avoid accusations of ideological ulterior motives, and so as not to give the military government a pretext for disqualifying them from politics.

In this context of deep political reconstruction, the civil society dynamics met a certain response in the authorities' action. It was, in particular, the case in Paraná, where "progressive" currents gained a foothold in the local administration and where some social concerns were included in environmental programs operating at that time. In particular, the defense of the "traditional" populations of family farmers and fishermen of the coast and of their local culture, known as *caiçara* (Adams 2000), was among the aims of the programs developed in the coastal area. Despite these social concerns, mistrust was still directed at these same populations because of their assumed negative impact on the Mata Atlântica, the last remnant of a highly valued natural patrimony. Thus, at a time when specific aid policies were targeted toward small family farmers of the interior high plains, no significant and sustained effort was made to understand the specific farming systems of the coastal populations, nor to bring agricultural extension support and special help to them (Miguel 1997).

Within this arena, nature conservation measures were put into effect by the new state government when it took office. In 1984, an ambitious program of environmental management was launched for the whole state. In 1985, the APA of Guaraqueçaba[2] was created. In addition, an agreement with the neighboring State of São Paulo was signed in order to harmonize measures taken on the same coastal zone on both sides of the border. Its declared objective was the preservation of the natural ecosystem and the culture of the traditional *caiçara* populations.

The explosive situation concerning land and social conditions described above figured among the concerns that inspired these interventions. Large landowners constituted the first targets of the control. However, in the absence of a specific policy of land reform, which today remains one of the major sources of conflict in Brazilian society, the land tenure problem could not be resolved. Even if large landowners were constrained in order to lessen their pressure on the forest, the structure of land ownership remained practically unchanged and they kept their possessions in the APA of Guaraqueçaba as long-term land investments; we will see that this calculation was justified. For the small or middle-sized farmers, the situation and the stakes were very different. They have had no choice but to remain where they were, to keep working their land and exploiting the surrounding natural resources, in order to secure their daily subsistence, and their futures, for themselves and their children.

Ecological Sanctuary or Development Zone: The Local Contradictions of Sustainable Development

As the most visible detrimental activities of the larger latifundia ceased, conflicts crystallized around a confrontation between small family farmers concerned for their future and outside institutions mainly animated by objectives directly or indirectly related to nature conservation. Among these outsiders, those coming from civil society organizations or representing private interests play an increasingly significant role. Available data indicate that, during the last few decades, the Mata Atlântica expanded appreciably in the *município* of Guaraqueçaba and more widely in the northern coastal area of Paraná (Marchioro 1999; Rodrigues 2002). This expansion is the combined result of environmental protection and a general decline in small family agriculture.

Even though the Mata Atlântica is no longer endangered in the coastal area of Paraná, it has now gained prominent symbolic status at the national and international levels. The Guaraqueçaba APA constitutes one of its main remnants. Since 1992, the whole region has been classified as a UNESCO Biosphere Reserve. As a result, there is increasing environmental control of the locality, with farmers being overwhelmed by regulations that severely restrict their productive activities. For example, a rigid prevention of forest clearance makes it almost impossible to continue shifting cultivation. These controls are in addition to the prohibitions on harvesting palm hearts and hunting and a strict limitation on woodcutting, even for personal use, including building and fence maintenance, and small handicraft production. This set of legal constraints adds to the difficulties caused by land degradation, which has affected most of these small farms, prior to the 1980s. Under these conditions, the very basis of their livelihood is threatened.

However, the landowners and users present in the APA are not equally affected by the current situation. Those having access to significant areas of arable land or forest are able to maintain profitable activities. In particular, they often benefit from subsidies for forest resource management, launched in the 1970s and reserved for properties of 3,000 hectares or more, which permit the continued exploitation of palm heart resources. Official authority often covers up for clandestine collection so that, even if the reliable data are lacking, alternative information leads us to think that the gathering, processing and marketing of palm hearts remains, even now, one of the most profitable sectors of resource exploitation in Guaraqueçaba.

As for the middle-sized farmers, most are newcomers who settled in the area only since the early 1980s. They subsequently engaged in rice growing, fruit cultivation, and operating plantations of exotic varieties of palm trees, all of which require more expensive techniques. These medium-sized enterprises can progressively modify their production techniques toward more ecologically safe processes, even benefiting from the increased commercial value brought to their products by these methods.

Finally, there are the small family farmers, who represent the bulk of the farming population of the *município*. These farmers are reduced to highly diversified, subsistence mixed farming, which scarcely enables them to survive. Under these conditions, many of them have no choice but to pursue activities that contravene the regulation of environmental protection, including field clearing in areas that, for the forest police, are difficult to reach, hunting, and, especially, the clandestine collection of palm hearts. This illicit harvesting can continue because it is beneficial to marketing networks traceable to local large landowners (Santos 2001). Many of these small-scale farmers are currently in a situation of both material and moral distress; they become the easiest and the most direct targets of repressive actions that are sometimes very violent (Zanoni et al. 2001).

Globalization of the Local Scene and Retreat of the State

As recent developments unfold, the situation is modified appreciably, introducing new elements in the relationships between local development and nature conservation. The protective measures, which are focused on the northern coastal zone of Paraná, translate the symbolic involvement that the State of Paraná and the nation of Brazil have made in this part of the forest. Indeed, Dean (1995) has shown how efforts to save the last remnants of the Mata Atlântica became a national campaign during the 1980s. In just a few years, the isolated, small, rural district of Guaraqueçaba was catapulted to the forefront of the national and international scene. During the 1990s, the most obvious sign of this change of scale was the increasingly significant position taken by new actors in the local arena. Neither civil servants nor local political representatives, these actors appear dependent, in a more or less direct way, on private interests with national and international linkages.

Thus, in 1997, a large Brazilian cosmetics company, O Boticário, acquired the lands of two old *fazendas,* including approximately 7,000 hectares of pastures and forest. The official classification as Reserva Particular do Patrimônio Natural (RPPN or Private Reserve of Natural Patrimony) was secured and, through an eponymous foundation, the company undertook to protect the existing forest and regenerate degraded areas. In addition, O Boticário devoted a significant part of its funds to the financing of research and awareness activities on the biodiversity of the Mata Atlântica. There is no evidence that actions in Guaraqueçaba, led by the Fundação O Boticário (Boticario Foundation), are directly driven by the cosmetics firm's interests; scientists retain a key role in the management of the reserve and the activities that take place there. However, one can question the model of development that inspires this initiative.

The reserve of O Boticário was situated in the vicinity of Morato, a community of the Guaraqueçaba *município.* For a long time, its inhabitants have exploited the resources of the surrounding forest and had been working in *fazendas* purchased by the foundation. The persons in charge of the reserve quickly realized that, if they wanted to avoid ceaseless conflicts with clandestine collectors of palm hearts and forest products, they needed to provide a substitute resource for former users of the forest. Thus was developed a handicraft production of baskets, designed to display the firm's products. The idea of "sustainable development," which is explicitly associated with this project by its promoters, is subject to the priority of preserving a natural patrimony with scientific, and possibly industrial, interests. Actions directed at the occupants and the users of protected areas are not primarily guided by concern for their current well-being and/or their future development but by the wish to control the use of rich, but threatened, ecosystems. Except for a few people who managed to benefit through their personal bond with the people in charge of the reserve, the majority of the Morato community inhabitants remain in a precarious relationship of dependency and survive under the most rudimentary material conditions.

A second new actor, the Sociedade de Pesquisa em Vida Selvagem e Educação Ambiental (SPVS – Society for Research on Wildlife and Environmental Education), will clearly affect the future of the northern coastal area of Paraná. This NGO was created in Paraná in 1984 by a group of academics who wished to deploy their expertise on behalf of nature conservation. Later, it obtained support from The Nature Conservancy, a North American organization. Until now, and along with an environmental education program, its activity within the APA has been centered on leading and financing research programs: documenting the flora and fauna, mapping land use, and carrying out ethnobotanical investigations and socioeconomic studies on local communities. The SPVS has been the main contributor of funds for research on Guaraqueçaba during the last decade and currently has, in Curitiba, the most complete documentation center in the *município*. In addition, it has engaged in some grounded development activities, particularly in the health field.

Boosted by its expert position, SPVS has carried out protection campaigns for large, threatened animals in Paraná. Furthermore, it gradually found itself taking on the role of advisor for the public agencies in charge of the protection of the Mata Atlântica, such as the Instituto Brasileiro do Meio Ambiente e dos Recursos Naturais Renováveis (IBAMA), or the Brazilian Ministry of the Environment and Renewable Natural Resources, and the Forest Police force. Thus, in Guaraqueçaba, this NGO gradually acquired a dominant position among the many actors playing a role in environmental protection. This position has been hegemonic since 1999, when the support of The Nature Conservancy brought to SPVS the possibility of playing a crucial role in the realization of a program for carbon sequestration in the APA, for the benefit of a U.S. electricity production company. Planned for forty years, this program is valued at 5.4 million dollars. Acting as project manager, SPVS carries out all the steps necessary for the acquisition of land, and its delimitation and classification as "Private Natural Reserves." It ensures the natural rehabilitation of land and the application of a program of controlled management of natural resources. Of a planned 20,000 hectares, 7,000 hectares have already been acquired. In addition, a land tenure survey was undertaken in areas likely to be directly or indirectly affected by the program. When public conservation areas are added, it shows the true area of land subjected to environmental protection and management (see Figure 2).

SPVS's developing relationship with international interests has radically changed its status in the local arena. From a position of influential protagonist, it has become the dominating actor, one holding the future direction of the area in the balance. Other public and private organizations acting in the area, including federal and state bodies in charge of environmental policy implementation,

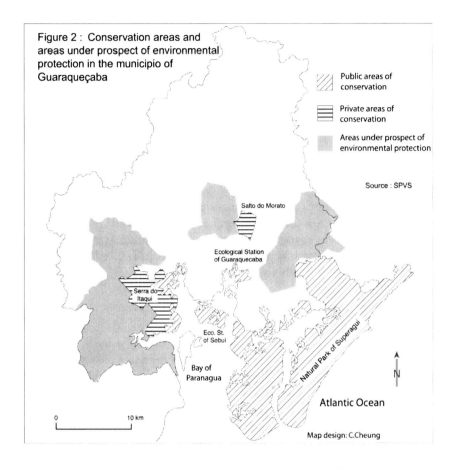

Figure 2 : Conservation areas and areas under prospect of environmental protection in the municipio of Guaraqueçaba

Public areas of conservation

Private areas of conservation

Areas under prospect of environmental protection

Source : SPVS

Salto do Morato

Ecological Station of Guaraquecaba

Serra do Itaqui

Eco. St. of Sebui

Bay of Paranagua

Natural Park of Superagui

Atlantic Ocean

0   10 km

N

Map design: C.Cheung

Figure 2. *Conservation Areas and Areas Under Prospect of Environmental Protection in the Município of Guaraqueçaba*

political and administrative institutions of the *município*, NGOs, and unions, are far from wielding the same level of power that SPVS draws from its international support and its immense financial resources.

This development meets little significant opposition at the local level. Small family producers complain and react by strategies of avoidance or dissimulation (Ferreira et al. 2002), but they do not succeed in organizing themselves to create the conditions for negotiation. The local municipality continues to follow a strategy whereby power is founded on the redistribution of financial resources in favor of partisans and managed to account for control of votes. A certain number of notables are directly or indirectly implicated in the principal economic activity of the zone: the exploitation of palm hearts. As this activity is largely clandestine in nature, they prefer not to engage in direct confrontation with the protectors of the environment. Finally, the owners of latifundia observe with a certain interest projects that might lead to an appreciation of the land capital they had accumulated three decades ago.

From now on powerful outside contributors, directly dependent on national or international private interests, have a free hand in defining priorities concerning the environment of Guaraqueçaba and its inhabitants. Such a situation raises a question of general policies related to the most fundamental options for the respective positions of the state, the affected populations, and the private sector. These policies define the future of the territory and the national community. With respect to the theoretical questions raised at the beginning of this paper, it also calls into question the definition of "sustainable development." When the dominant actors "capture" such a notion, their main concern is not to promote the development of the local community. It is, in fact, to reduce the social strains that are prejudicial to the local environment, by substituting resources previously drawn from the forest with new ones. "Social sustainability" becomes subservient to "natural sustainability" and not the reverse.

Conclusion

In returning to the theoretical considerations raised at the beginning of this essay, we can see how much recourse to history is necessary to reexamine, in a precise local context, the concept of sustainable development. Indeed, what are the lessons that we can draw from the study of the Guaraqueçaba situation? This case confirms that the relation between the forest and its resident populations has always been marked by change. The Mata Atlântica has indeed undergone formidable challenges during the past century. One such challenge occurred during the first two decades of the twentieth century, when banana cultivation was in full development. The second took place during the 1960s and 1970s, under the influence of public policies that promoted the influx of new economic actors guided by the objective of quick profit. However, two factors have ensured the area's relative safety. First, specific geographical conditions in the northern coast, including topography, isolation, and climate, placed the zone in an unfavorable competitive position in comparison with the high plains areas, where the essential efforts of agricultural intensification have been concentrated until today. Second, the power acquired by environmental interests during the period toward the end of the dictatorship has led to an end to the forest destruction that was begun by logging companies and the owners of latifundia.

In the course of its history, the forest has shown resilience. Studies of land use evolution show that the forest has recovered ground to a significant degree during the last twenty years as the result of environmental policies and as a consequence of changes in the agricultural system. The Mata Atlântica bears no resemblance to an image of "primary" or "virgin" rain forest, either now or one hundred years ago. It is the product of a natural and social history but remains a specific forest ecosystem with ties to the Mata Atlântica of centuries ago.

Local society has also undergone profound transformations. It was marked during the last two centuries by migrations of great scope, which produced a population mix that continues to this day. The "traditional" image of *caiçara* society no longer conforms to present reality. The feeling of belonging to a place and a community remains strong among some families, but this has been subjected to continuous rebuilding, accentuated by the arrival of many churches and sects that have met with great success among the local population. In addition, the expansion of a market economy has created new forms of economic and social differentiation. There is now a strong contrast between the majority of farming families whose livelihoods came from subsistence production, and a few farms with developed strategies of land accumulation and modernized instruments of production, for which the bulk of production is intended for the market.

Today, local small farmers are severely threatened both in their material survival and in their capacity to maintain an enduring sense of identity. Many of the younger generation migrate to the city and gradually become integrated into another social and cultural universe. If "social sustainability" consists not only of biological reproduction but also of the capacity to project personal and community identity into the future with continuity of affiliation and feelings of membership, many families and local groups in Guaraqueçaba are deeply threatened in their social existence. Evolution of the rural world, in Brazil as in many other places around the globe, leaves less space for family farming. The process has been worsened by the symbolic meanings associated with the defense of the Atlantic forest, and its flora and fauna; humans are often relegated to the status of unwelcome visitors. "We are not at home here and we count less than the *micoleão* monkey or the parrot!" is the complaint often heard. The paradox is that, although their traditional farming practices have never been intrinsically destructive of the forest[3], today these practices constitute the principal targets for corrective action. The case of Guaraqueçaba offers a particularly eloquent illustration of the difficulty in reconciling two conceptions of sustainable development: one that privileges natural sustainability and one that privileges social sustainability.

The major dynamic feature in the northern coast's recent history is the change regarding the scale of political, social, and economic space within which it develops interactions. Having remained for decades a small local space loosely articulated to the general changes of Brazilian rural society, Guaraqueçaba has suddenly become a focal point in a network of interests that are played out at regional, national, and international levels. This is a consequence of the value of its natural environment, today considered as belonging to a common world patrimony.

The analytical "grid" presented at the beginning of this essay, which situates local actors' representations of sustainable development according to an opposition between the "here" and the "elsewhere" and the "present" and the "future," provides us with a powerful tool of analysis. Most of the farmers within the local population think of their sustainability in relation to the local habitat in which their lives are played out. They conceive of theirs and their children's futures in direct relationship to the way in which they can build their "present" in this particular context. For regional or national environmentalist movements, the case of Guaraqueçaba was and remains a significant though partial element at the service of a political strategy of positioning and advocacy, which has vastly exceeded the importance of any particular situation. For these actors, the primary goal is to dispute a model of development, as well as a form of society and culture, and to defend a new way of conceiving the relationship between human beings and nature. Intervening in a particular "place" is a way to pursue goals at a national level and to defend universal and timeless values. With the intervention of actors who move in the sphere of debates and the conflict of interests that encompass the question of global change, the concept of sustainable development moves to a different spatial and temporal scale. According to such a perspective, Mata Atlântica becomes an invaluable element of the world inheritance. Populations that either live there or exploit it are responsible for its preservation, and accountable before the rest of the planet, both for the present and for the distant future.

This case study offers an illustration of the differences and contradictions between various conceptions of sustainable development applied at a local level. Highlighted are the multiple perspectives of the different actors, which coalesce around the concerns of development and sustainability. For the actors, these issues are not conceptual nor speculative but, in fact, are positions that guide strategies of concrete action. Small farmers try to secure their material survival, but they also try to reconstruct their legitimacy as occupants of the place by appropriating the speech of recent ecological movements. They present themselves as primary defenders of nature, even as they are obliged to practice activities that contravene the regulations in force. Farmers coming from a zone that takes advantage of a strong symbolic investment could enter the market and begin to benefit from an appreciation of organic agriculture production. Large landowners seek to realize increased value for their land because of its patrimonial importance.

NGOs, such as the O Boticário Foundation and the SPVS, concentrate diverse interests and objectives within their organizations. Both scientists and environmentalists, who may still remain faithful to the idealized vision of nature that animates the ecologists, and managers who are more inclined to favor the institutional objectives and the requirements of their silent partners, even if it may create conflict with their initial ideals. Then there are the external protagonists, who often exert considerable influence on local-scale events, who have to show a profit, at least indirectly, on the investments they granted to Guaraqueçaba. Political leaders can be added to the list of stakeholders: they are concerned with creating alliances between some of the groups involved in this process and, in particular, maintaining the popular base necessary for their re-election. In this arena, the role of public organizations representing the state, at both the regional and federal levels, is increasingly discrete.

This history of the northern coastal zone of Paraná provides elements inform-ing a theoretical reflection on the analysis of the society-nature relationship. More precisely, the sustainable development concept, as applied at the local scale and under the flag of nature conservation, is informed by this history. We would like to emphasize that the concept of stability–of static balance–does not offer an operational tool to analyze the relationships that, within a local space, are linked over time between a population and its environment. The relationships at stake always contain tensions of a dialectical nature. The history of the past century shows that, in Guaraqueçaba, the changes that occurred became perma-nent and led to reciprocal adjustments of the natural environment and local society. The forest we observe today, with its high level of resilience, is the prod-uct of this history. Preserving the current Mata Atlântica does not mean the protection of a relic; rather, it means dealing with the forest as a fragment of his-tory. The social scene has also undergone radical changes. The current rural society has little in common with that which existed at the first half of the last century, even if the thread of identity is not completely broken in certain com-munities. Moreover, new agents, arriving from outside the locality and driven by different logics and interests, gradually make their appearance and introduce new interests. The notion of "ethnoconservation" clearly cannot apply to such a dynamic social reality.

The conditions of articulation between local realities and external spheres, between the "here" and the "elsewhere," constitute the pivot of this dynamic of transformation. These conditions include the wave of expansion and then collapse of the banana market; ceaseless demographic flows in and out of the region; the impact of public policies that involved new economic agents; local nature reconstituted as part of common patrimony under the regional and national emergence of "Green" movements; and its positioning on the inter-national stage and its subsequent connection to global interests. These dra-matic events have not only followed one another but, to some extent, they were stratified so that the actors corresponding to each of them are still pres-ent. Each one has objectives and interests, its own vision of development and interpretation of sustainability. Consequently, the question of "sustainable development for whom?" varies according to this diversity of positions and views, a diversity that is largely determined by the position occupied with respect to the local arena.

Notes

1. For a detailed reconstruction of the agrarian history of coastal northern Paraná, see Miguel (1997:27).
2. The APA covers a territory of 3,134 km2, which includes the total *município* of Guaraqueçaba (2,317 km2 and approximately 8,000 inhabitants) and some portions of Antonina and Campina Grande do Sul *municípios* (see Figure 1).
3. They became destructive only under the effect of external constraints: the rejection of the poorest on the mountainsides during the banana boom and the land erosion leading to the reduction of regrowth time under the effect of large latifundia. In the organization of palm heart collection, small farmers are only an ultimate link in a network that is firmly organized and directed by powerful patrons who are rarely concerned (Laulan 2002).

Acknowledgments

We wish to thank Christopher Taylor, Phil Bradley, and Barbara Yablon Maida for their help in revising the original English manuscript of this text.

References

Adams, Cristina. 2000. *Caiçaras na Mata Atlântica: Pesquisa científica versus planejamento ambiental.* São Paulo: Anablume/FAPESP.

Amselle, Jean-Loup. 1990. Logiques métisses. Anthropologie de l'identité en Afrique et ailleurs. Paris: Payot.

Balandier, Georges. 1989. *Le désordre, éloge du mouvement.* Paris: Fayard.

Beckerman, Wilfred. 1972. *Economists, Scientists and Environment Catastrophe.* Oxford: Oxford Economic Papers.

Chang, Man Yu. 2000. *Seqüestro de carbono florestal. Oportunidades e riscos para o Brasil.* Mimeo, Doutorado em Meio Ambiente, UFPR, Curitiba.

Dean, Warren. 1995. *With Broadax and Firebrand: The Destruction of the Brazilian Atlantic Forest.* Berkeley: University of California Press.

DeAngelis, Donald L., and J. C. Waterhouse. 1987. "Equilibrium and Monoequilibrium Concept in Ecological Models." *Ecological Monographs* 57, (1): 1-21.

Diegues, Antonio C. 1996. *O Mito Moderno da Natureza Intocada.* São Paulo: Hucitec.

___, ed. 2000., *Etnoconservação. Novos rumos para a proteção da natureza nos trópicos.* São Paulo: Hucitec.

Dubois, Jean-Luc, and François Regis Mahieu. 2002. "La dimension sociale du développement humain durable: réduction de la pauvreté ou durabilité sociale?" In *Développement durable? Doctrines, Pratiques, Évaluations,* ed. Jean-Yves Martin. Paris: IRD (L'Institut de Recherche pour le Développement) Editions.

Elame, Esoh. 2001. "Repenser le concept de développement durable." *Cahiers du G.R.A.T.I.C.E.* Université Paris XII-Val de Marne, no. 20: 135-53.

Ferreira, Angela, Eduardo B. Santos, Catherine Dumora, and Edna C. Francisco. 2002. *O ocultamento como estratégia de reprodução dos agricultores familiares em Guaraqueçaba.* Congresso da ALASRU, Porto Alegre.

Holling, Crawford S. 1976. "Resilience and Stability in Ecosystems." In *Evolution and Consciousness: Human Systems in Transition,* ed. Erich Jantsch and C. H. Waddington. London: Addison-Wesley.

IUCN (The World Conservation Union). 1980. *World Conservation Strategy.* Gland, Switzerland.

Jacobs, Peter, and David Munro, eds. 1987. *Sustainable and Equitable Development: An Emerging Paradigm.* Geneva: IUCN (The World Conservation Union).

Jollivet, Marcel, ed. 2001. *Le développement durable, de l'utopie au concept.* Coll. Environnement, Paris: Ed. Elsevier/NSS.

Larrère, Catherine. 1997. *Les philosophies de l'environnement.* Paris: Presses Universitaires de France.

Latouche, Serge. 1995. *La mégamachine. Raison scientifique, raison économique et mythe du progrès.* Paris: La Découverte.

Laulan, P. 2002. "Dynamique spatio-temporelle de la Mata Atlântica (Batuva, commune de Guaraqueçaba, Etat du Parana, Brésil): Evolution des modes d'usages et transformation du couvert forestier de 1953 à nos jours." Dissertation, DEA Environnement: Milieux, Techniques, Sociétés, Museum National d'Histoire Naturelle, Paris.

Little, Paul E. 1999. "Environments and Environmentalisms in Anthropological Research: Facing a New Millennium." *Annual Review of Anthropology* 28: 253-84.

Ludwig, Donald, Brian Walker, and Crawford S. Holling. 1997. "Sustainability, Stability and Resilience." *Conservation Ecology* 1 (7), .

Marchioro, N. P. 1999. "A sustentabilidade nos sistemas agrários no litoral do Paraná ; O Caso de Morretes." Dissertation, Meio Ambiente e Desenvolvimento, UFPR, Curitiba.

Martin, Jean-Yves, ed. 2002. *Développement durable? Doctrines, Pratiques, Evaluations.* Paris: IRD (L'Institut de Recherche pour le Développement) Editions.

Miguel, L. A. 1997. "Formation, évolution et transformation d'un système agraire dans le sud du Brésil (littoral nord de l'Etat du Paraná): une paysannerie face à une politique de protection de l'environnement: Chronique d'une mort annoncée. " Dissertation, Institut National Agronomique, Paris-Grignon.

ONU/EPHE. 1972. *Development and Environment.* La Haye–Paris: Mouton.

Perroux, François. 1961. *L'économie du 20ème siècle.* Paris: Presses Universitaires de France.

Pezzey, John. 1989. *Definitions of Sustainability.* London: CEED.

Pimm, Stuart L. 1991. *The Balance of Nature.* Chicago: University of Chicago Press.

Rodrigues, A. S. 2002. "A sustentabilidade da agricultura em Guarequeçaba: o caso da produção vegetal." Dissertation, Meio Ambiente e Desenvolvimento, UFPR, Curitiba.

Rolston, Holmes. 1994. *Conserving Natural Value.* New York: Columbia University Press.

Rossi, Georges. 2001. *L'ingérence écologique. Environnement et développement rural du Nord au Sud.* Paris: CNRS (Centre National de la Recherche Scientifique).

Sachs, Ignacy. 1990. *Stratégies de l'écodéveloppement.* Paris: Les Editions Ouvrières.

Santos, E. B. 2001. "Estratégias dos agricultores familiares em uma comunidade de Guaraqueçaba." Dissertation, Curso de Ciências sociais, UFPR, Curitiba.

van der Leeuw, Sander E. 1998. "Archeomedes, un programme de recherches européen sur la désertification et la dégradation des sols." *Natures, Sciences, Société* 6 (4) : 53-58.

WCED (World Commission on Environment and Development). 1987. *Our Common Future.* Oxford: Oxford University Press.

Zanoni, Magda, Lovois de A. Miguel, Angela D. Ferreira, et al. 2001. "Protection de la nature et développement rural. Dilemmes et stratégies des agriculteurs familiaux dans des Zones d'Environnement Protégé." In *Agriculture et ruralité au Brésil*, ed. Magda Zanoni, Angela Ferreira, et al. Paris: Karthala.

Alaska Native Corporations and Subsistence
Paradoxical Forces in the Making of Sustainable Communities

Thomas F. Thornton

Introduction

In America's largest state, Alaska, the environment is not an issue but *the* issue. This is clearly reflected in the state's conflicted material and symbolic status as "the last frontier," which brands it as both an unspoiled wilderness for preservation and a land of vast untapped resources for industrial development. Somewhere between the wilderness preservationists, whose influence has helped make Alaska a land of parks, preserves, and tourism, and the industrial developers, whose power has shaped Alaska's natural resource-based economy, dwell Alaska Natives, who historically have sought sustainable relationships with the lands they have inhabited for centuries, if not millennia. Today, Alaska Native economies are pushed from both sides.

On the one hand, Alaska Natives have been transformed into "corporate Indians," shareholders in business corporations created by the U.S. Congress to help them manage real estate and monies awarded through land claims. On the other hand, Alaska Native rural, noncommercial or subsistence economies are also protected by federal law. When the state's Native (Indian, Eskimo, and Aleut) communities were reengineered from small-scale subsistence-oriented economies into for-profit corporations by the landmark Alaska Native Claims Settlement Act (ANCSA) of 1971, many predicted that corporate capitalism would swallow their cultures and assimilate them into the mainstream American economy and society. Thirty years after the advent of this experiment, we are in a better position to assess this prediction. Indeed, ANCSA has been an unprecedented experiment not only in social engineering among aboriginal peoples but also in corporate engineering, for ANCSA corporations differ from common capitalist corporations in important ways.

This essay examines the legacy of ANCSA within the context of globalization, and specifically whether the unique corporate model and culture that has been forged in Alaska offers an alternative to the increasingly deracinated, competitive, short-sighted, and often corrupt corporate culture that is developing transnationally. How have ANCSA corporations reshaped Alaska Native cultures, and, in turn, to what extent have Alaska Natives remolded their corporations to be more congruent with their cultural ideals and ways of being? Among the most important of these ideals is the maintenance of subsistence lifeways in Native communities. The seemingly paradoxical forces of "progressive" capitalist Native corporations spawned by federal law to foster industrial development, on the one hand, and federal subsistence policy that protects customary and traditional subsistence uses of fish, wildlife, and other local resources, on the other, have fostered a contentious debate within Native communities over environmental, economic, and social policy. Underlying this debate is a question of broad significance for Alaska and beyond: Can modern business corporations and traditional rural economies be integrated to create sustainable communities?

Background

The Alaska Native Claims Settlement Act of 1971 (1971, 43 U.S.C. & SECT; 1601 et seq) was the culmination of many years of wrangling between the federal government, the state of Alaska (admitted to the Union in 1959), and Alaska Natives over how to settle Native claims to virtually all of the state's roughly 400 million acres. Alaska Natives are unique among Native Americans in having never been confined to reservations (with the exception of the Tsimshians of Metlakatla) and having remained relatively undisturbed through much of the historical period when Indians in the "Lower 48" were being dispossessed and subjugated. Despite contact with the Russians dating to the late eighteenth century, followed by European and American fur traders, Alaska Natives remained largely autonomous over their lands until the late nineteenth century. But after the sale of the Alaskan territory to the United States in 1867 for $7.2 million, Alaska's first peoples suffered waves of incursions by commercial fishing, mining, timber, and other interests. Notwithstanding appeals, legislation, and land claims commissions, Natives were unable to secure meaningful legal protections or title their traditional lands and resources in the face of this onslaught. This was due not only to their lack of political enfranchisement, as U.S. citizenship came only in 1924, but also to powerful corporate interests that conspired with government to limit Native claims.

Finally, in the late 1960s the impetus for a settlement arrived with the discovery of huge quantities of crude oil on Alaska's North Slope. To exploit the oil it would be necessary to develop a vast area and infrastructure, including a trans-Alaskan pipeline to move the crude from the frozen northern tundra to the southern port of Valdez, where it could be shipped to markets outside the state. Competing claims to the lands and resources in the region encompassing oil development had to be removed before the government and industry could move forward with their business plan. Whereas in the past corporate interests had always opposed Native claims to lands and resources, oil companies at least now favored a claims settlement as a way of limiting their liability and "opening up" the land without further encumbrance. Thus, ANCSA was pushed through Congress with little opposition (Mitchell 2001).

The law was much more than a real estate settlement, however. ANCSA radically reorganized Alaska's hunting and gathering peoples into 13 regional and 220 village for-profit corporations and invested them with significant capital, that is, money, land, and resource rights, in hopes that they would develop self-sufficiency. In terms of federal Indian policy, the law represented a mix of termination and self-determination themes. Federal termination policies of the 1950s had worked to remove tribes from their dependency relationship with the U.S. government by severing the government's responsibilities for many aspects of their social welfare. In the 1960s and 1970s, termination was abandoned in favor of policies designed to increase tribal self-determination and control over their lands and resources. But most tribal governments lacked the capital, infrastructure, and cultural ethos to develop their lands, and legally the U.S. government continued to have a trust responsibility over Native Americans. ANCSA corporations were devised as an institutional means to give the tribes the capital, infrastructure, and incentives to develop and manage their own natural assets, thereby increasing their self-sufficiency and reducing the government's trust responsibility.

Major structural provisions of ANCSA included the following:

- "Congress finds and declares (a) there is an immediate need for a fair and just settlement of all claims by Natives and Native groups of Alaska, based on aboriginal land claims; (b) the settlement should be accomplished rapidly, with certainty, in conformity with the real economic and social needs of Natives, without litigation, with maximum participation by Natives in decisions affecting their rights and property, without establishing any permanent racially defined institutions, rights, privileges, or obligations, without creating a reservation system or lengthy wardship or trusteeship…" This language neatly encapsulates the termination and self-determination threads within ANCSA.
- Aboriginal land title is extinguished, in effect eliminating Indian Country status in the state and the sovereignty inherent in that status.

- Aboriginal hunting and fishing rights are extinguished. The continuing subsistence needs of Alaska Natives are recognized in the legislation, however, and are to be protected by the Secretary of the Interior and the State of Alaska.
- Natives are awarded $962.5 million in compensation for lands lost. This money is not distributed directly to individuals or tribes but put under the control of newly established ANCSA corporations (see below). The compensation amounts to roughly $3 per acre of land.
- Natives are given title to roughly 10 percent of Alaska (44 million acres). In addition to compensation for lands taken, Natives also receive title to a portion of their traditional lands through the newly created corporations. The federal government retains control over 60 percent of Alaska lands and the state controls the remaining 30 percent. Of the 44 million acres conveyed, 22 million acres of surface estate was apportioned to village corporations on the basis of population. For the most part, this land surrounds the villages and consists of prime subsistence and natural resource areas. The subsurface estate of this land was conveyed to the regional corporations, along with 16 million acres of separate land for development and 2 million acres of land identified as historical sites and cemeteries.
- Thirteen regional and 220 village corporations are established to control settlement lands and money. Eligible Natives become shareholders in village and regional corporations. Most receive 100 shares of stock. Those born after 1971 receive nothing, except through inheritance or other transfer.

For many Natives, the single most important outcome of ANCSA was increased control over lands. Willie Hensley (2001), an Inupiat Eskimo leader involved in land claims negotiations, summed it up this way:

> The reality was that we changed United States policy 180 degrees. They [the U.S. government] had been taking land for 200 years, and all of a sudden they were finally conveying. Even if it was 40 million acres, that's still 40 million acres. It was done in a way that gave the people the right to make decisions about what happened to it, and that was a major, major change.
>
> …Up until that time [1971], the missionaries and the government were all-powerful. They were in control of key institutions that had effects on peoples' lives. We didn't have any say about the most elementary things that affected our families. From that standpoint, I think the Settlement could affect not only the lives of our communities, but also the entire direction of the state. Alaska itself will never be the same again as a consequence of the Native peoples' efforts to control the lands they had lived upon.

Indeed, the signature slogan of the land claims movement was "Take our land, take our life." And in the early post-settlement phase, young Native leaders recall the advice of elders, who cautioned: "Whatever comes of this effort, its purpose is to protect native lands for future Native generations, to keep the land for us" (Mallott 2001b).

But critics of ANCSA were skeptical, including the distinguished Canadian jurist Thomas Berger (1991: 104-5), who views the allocation of land under ANCSA corporations essentially as "[a]nother version of the General Allotment Act" of 1887 in that it abrogated traditional communal ownership and prerogatives over lands and resources in favor of a Western model of progressive land exploitation. "At each stage the United States wanted its Native population to adopt the current paradigm of progress: in 1887 it was farming, in the 1950s it was factory work, in 1971 it was business."

Some Alaska Natives, especially those leaders with Western educations, accepted ANCSA's business model of modernization, development, and acculturation as inevitable, if not the most favorable way for Natives to gain control over their futures. The rationale for embracing ANCSA was that Native-run corporations would protect ANCSA lands for Native use and serve as a basis for achieving financial prosperity within the dominant capitalist society. Profits derived from corporations would, in turn, be used to improve conditions in Native villages and communities. It seemed a practical recipe for building sustainable Native communities.

Yet many Natives were opposed to the corporate model from the outset because they feared it would take land and other assets away from traditional governing institutions of the tribes and put them in the hands of business organizations bent on exploiting and alienating them as commodities. Similarly, many worried that the corporate economy would undermine the noncommercial subsistence economy by depleting the natural resources on which Natives had depended for centuries. In fact, this was already happening in industrialized Alaska, where the seemingly inexhaustible salmon and other marketable fish and wildlife resources had become overexploited. Since the late nineteenth century, Natives had petitioned the government for provisions to protect their natural resource base and subsistence economies in order to ensure their sustainability, but to no avail (Goldschmidt & Haas 1998).

Despite ANCSA's termination of aboriginal hunting and fishing rights, the law provided for "both the Secretary [of the Interior] and the State [of Alaska] to take any action necessary to protect the subsistence needs of Natives." This rather vague and paternalistic clause turned out to be an Achilles' heel in the legislation for Natives in that it provided no immediate protection or priorities for subsistence resources or subsistence economies. As Hensley (2001) recalled, "we couldn't get the subsistence issue resolved, initially. We came back to try to deal with it, but once you get involved with the political system like that, the outcome is not altogether sure, because what power do we have?" A legislative solution to the subsistence problem did not come about for another decade, when Congress produced a federal subsistence law, which became the basis for regulating Alaskan subsistence economies.

The heart of federal subsistence law is Title VIII of the Alaska National Interest Lands Conservation Act (ANILCA) of 1980 (PL 96-487). This law gave allocation priority to subsistence uses of wild resources over other consumptive uses, such as recreational hunting and commercial fishing, in times of shortage. ANILCA, like ANCSA, was the product of tenuous political compromises between competing interests. As such, it has been pivotal in framing the contemporary subsistence problem in terms of three divisive cleavages: (1) a cultural divide between Natives and non-Natives; (2) a rural-urban split in allocation of scarce resources; and (3) a federal vs. state conflict over management authority (Thornton 1998, 2001).

Federal law makes no attempt to define subsistence itself but only "subsistence uses." These are "the customary and traditional uses by rural Alaska residents of wild, renewable resources for direct personal or family consumption" (Sec. 803). Yet the law does recognize a qualitative difference between Native and non-Native subsistence, wherein "the continuation of the opportunity for subsistence uses by rural residents of Alaska, including both Natives and non-Natives … is essential to Native physical, economic, traditional, and cultural existence" and "to non-Native physical, economic, traditional, and social existence" (Sec. 801). The differences between "cultural" and "social" are not specified, but the distinction does make clear that Alaska Native subsistence practices are enduring and fundamental to their cultural survival. Ideologically, the distinction is emblematic of Natives' more expansive and foundational definitions of subsistence, often expressed in aboriginal languages as "our way of living," as opposed to non-Natives' more minimalist, economic definitions of subsistence as basic sustenance (Thornton 1998).

However, ANILCA's recognition of the centrality of subsistence to Alaska Natives' cultural existence was not enough to guarantee them priority, much less exclusive, rights under federal law. Through a political compromise, ANILCA awarded an allocation preference on the basis of rural residency rather than ethnicity, even though ethnicity-based preferences had earlier been granted for Alaska Native coastal community residents through the Marine Mammal Protection Act of 1972. In the case of fish and game species for which non-Native sport and commercial interests compete, however, the stakes were too high, and the state and other non-Native interest groups vigorously opposed any form of Native preference or collective rights. At first glance, the rural preference seemed to at least partially fulfill ANSCA's promise to provide for the Alaska Native subsistence needs. After all, prior to 1980, the majority of Alaska's rural residents were Natives and, though many Natives migrated to urban centers, the majority still lived in rural communities that were dependent on wild resources.[1] But any long-range demographic analysis would have shown that these fragile majorities would not hold, and indeed they have not. Due to large migrations of non-Natives into the state, Alaska Natives by 1990 comprised less than 20 percent of the state's population and had become minorities in both rural and urban areas. Today more than half of Alaska Natives reside in urban areas, rendering them ineligible for subsistence harvests under ANILCA. At the same time, the growing non-Native urban majority in the state—three- quarters of Alaskans live in urban areas, about half in greater Anchorage—has become increasingly effective in attacking the rural preference at the state level (Thornton 1999).

On balance, then, the ANSCA/ANILCA solution of extinguishing Alaska Native hunting and fishing rights in favor of weak subsistence lifestyle protections under a dubious rural preference has created as many problems as it has solved. Despite lofty expectations, paradoxes and contradictions inherent in the ANCSA-ANILCA body of Indian law have served to divide interests along economic, cultural, geographic, and political lines. In addition, it has changed the socioeconomic and political organization of Alaska Native communities in important ways. Former Sealaska Corporation CEO Byron Mallott (2001a) notes, "ANCSA overran our leadership. … It parked the elders on the sidelines in some powerful ways. We paid homage to them, and we used them and listened to them when appropriate, but the exigencies and urgencies of making ANCSA work essentially ran over everybody, including ourselves." The subordinate status of subsistence to commercial and recreational economies seems inimical to building the long-term foundation for Alaska Natives' economic and cultural well-being that ANILCA sought. As Berger (1985: 65) observes: "In effect ANILCA is a partial restoration of Native hunting and fishing rights [eliminated by ANCSA], but it does not go far enough. More is required if subsistence is to remain a permanent feature of Native life and culture."

Studies show that rural subsistence accounts for only about 4 percent of the total fish and wildlife harvest, while 95 percent goes to commercial interests and 1 percent to sport users. Looking at these figures, one might wonder why there is such a fuss over subsistence, since users take such a tiny slice of the resource pie. On the other hand, subsistence remains a critical base in contemporary rural economies. The distinguishing feature of subsistence-based socioeconomic systems is the primary economic, social, and cultural reliance on fish and game resources. Cash and current technologies are utilized, but they are integrated into the community's economic and social activities "so as to be mutually supportive" (Wolfe & Ellanna 1983: 252). This stands in stark contrast to market-based societies in which the market sector is the cog of the economic and social organization. While no community in Alaska today is completely self-sufficient—all participate in mixed economies involving production for home use and consumption of market goods—rural communities rely on subsistence resources to meet their basic needs.

Currently some 270 communities, embodying roughly 21 percent of Alaska's 675,000 residents and 48 percent of the Native population, are classified as rural under federal law. Three basic characteristics of these rural Alaska subsistence economies are high harvest levels, reliance on kin groups for production and exchange, and traditional knowledge and resource tenure (Wolfe 1998). Annual harvest levels in rural Alaska range from 153 to 664 edible pounds per capita and are on average more than ten times higher than those in urban communities, like Anchorage, which range from 16 to 40 pounds per capita. The major components of the rural harvest include fish (60 percent), land mammals (20 percent), and marine mammals (14 percent), along with a wide range of invertebrates, birds, and plants.

Reliance on kin groups for labor and exchange is a form of economic organization sometimes termed the "domestic mode" of production. This mode of production lends itself to certain patterns of harvest and exchange. For example, households that include healthy, knowledgeable, and well-equipped harvesters will often specialize and harvest foods for their relatives and others. Thus we find consistently in Alaska Native rural communities that about 30 percent of the households produce about 70 percent of the wild foods. These high- producing "super-households" then distribute these resources through kinship and other social networks to family members and others in need, such as the elderly. In this system of generalized reciprocity, which is common in hunting and gathering societies, harvesters may receive material goods or social rewards in exchange for their gifts of foods, but this is not expected. These harvest and distribution patterns prevail despite the fact that fish and game regulations, such as bag limits, are geared overwhelmingly toward the *individual* needs of recreational users rather than the *communal* needs of subsistence users in the rural socioeconomic system.

Finally, traditional knowledge, or traditional environmental/ecological knowledge, and resource tenure play an important role in rural subsistence economies. Most Native subsistence users prefer to hunt, fish, and gather in resource areas where they have ancestral ties and intimate local knowledge. Customary rules and traditions concerning access and use and, as we have seen above, distribution of particular resources and resource areas continue to exist alongside state and federal regulations. Although neither federal nor state subsistence laws contain provisions to protect ties to place, as only customary and traditional uses are safeguarded, Native social organization and land and resource tenure systems maintain fidelity to place and limit demand on key resources.

As the above analysis reveals, subsistence economies are not—as another popular misconception holds—simple or antiquated lifestyles. On the contrary, subsistence activities are a vibrant and vital sector of the contemporary rural economy, providing basic nutrition, gainful employment (as subsistence is by far the state's largest "employer"), and valued products that would cost between $250 and 400 million to replace. These economic benefits, along with the host of positive sociocultural values engendered by subsistence, greatly increase the quality of life in rural Alaska. Nor are subsistence economies and the subsistence priority in any way a threat to the commercial economy, as competing interests, such as sport and commercial hunting and fishing businesses eager to limit subsistence protections, hold. On the contrary, it is the commercial economy that threatens subsistence (Thornton 1999).

Sustainability and the Idea of the Corporation

Can Alaska Native communities sustain themselves within the framework of a dual corporate-commercial and subsistence economic framework? If the guiding myth and logic of capitalist development and assimilation behind ANCSA is correct, Alaska Native corporations and the commercial economy will eventually triumph over subsistence economies. If the guiding myth behind ANILCA is correct, the two economies will continue to coexist in a mutually tolerant, if not supportive and synergistic way. Alaska Natives involved in the corporate economy typically pledge allegiance to the ANILCA subsistence mythology, but many village Natives, marginalized from the wealth of the corporate economy (see Dombrowski 2002), see their subsistence way of life threatened by the ANCSA corporate model. As one Tlingit Indian villager complained, she must cover her eyes when she flies over her community because it pains her so to see the massive destruction of wildlife habitat caused by her own ANCSA corporation's clear-cutting of timber, and she feels powerless to do anything about it (DJ, personal communication, 2001). On the other hand, studies show that subsistence fish and wildlife harvest levels in rural communities, despite the impacts of development, have not dropped precipitously since 1971 (ADF&G). In other words, the subsistence economy shows few signs of succumbing to the corporate one.

Thus, despite thirty years of coevolution between the two economies, the results of this experiment are still uncertain. This is partly because it is not a simple zero-sum game in which one economy triumphs at the expense of another, but rather a complex and dynamic political economy in which social and ideological forces also play a role. As key engines and anchors of the Alaskan commercial economy, Alaska Native corporations potentially can play a major role in shaping a sustainable future for both corporate and subsistence interests. But in order to succeed, they must become not simply economic institutions but viable sociocultural institutions as well. In a nutshell, this means reconceptualizing the very idea of the corporation within an increasingly anti-indigenous global economy. The provisions of ANCSA did not equip Native corporations especially well to do this; instead it has been a slow, uneven process of learning to balance competing interests that is still very much ongoing.

Table 1. Contrasts between Alaska Native and Conventional Corporations

Other for-profit Corporations	Alaska Native corporations (ANCSA)
Usually formed with plan to make profits. Membership and assets not linked to culture, ethnicity, or homeland.	Formed by Congress in 1971 to settle Native land claims and provide a vehicle for economic development, investment, and earnings.
Shareholders can make money by selling their stock or earning dividends.	Limited stock transfer: Shareholders not allowed to sell stock, unless corporation votes to approve stock sales. Dividends are paid.
Shareholders expect benefits in the form of dividends and growth in stock values.	Noneconomic benefits: Native corporations may provide social and educational benefits, which shareholders have come to expect.
Corporations make money for themselves.	Corporate socialism: Twelve regional corporations must share revenue from timber and minerals with each other and with 170+ village corporations.
Anyone owning shares can vote for directors.	Ethnic corporations: Shareholders must be Native or descendants of Natives to vote for directors.
As of 1986, lost the net operating loss (NOL) tax break whereby NOLs could be sold to profitable companies seeking tax relief.	Corporate welfare: ANCSA corporations could continue selling NOLs from 1986 to 1988, allowing more "breathing room" for those struggling.
Typically own large tracts of land only for commercial development. Land usually taxed.	Limited taxation: Hold millions of acres of land not only for development but also to protect subsistence, historic sites, and other cultural resources. Not taxed unless land is developed.
May go bankrupt, dissolve, and liquidate lands and other assets if unprofitable.	Corporations may go bankrupt, but thus far have not been forced to dissolve or liquidate lands.
Flexible accumulation: Corporations move capital where materials, labor, and other costs are lowest.	Rooted capitalism: Corporations face strong pressures to invest in their own communities.

In examining the evolution of ANCSA corporations, let us first compare the key features of ANCSA's unique ethnic corporate institutions to those of conventional U.S. corporations (Table 1). As the table shows, Alaska Native corporations are distinguished not only by their ethnic composition but also by their rootedness, redistribution of profits, and investment strategies.

ANCSA corporations did not replace tribal governments or traditional social, economic, and political structures. Rather, the regional and village corporations constituted two additional, dynamic layers on the existing structures (Thornton 2002). The result has been a fascinating, often conflicting, and perhaps revolutionary mix of the traditional and the modern, the local and the global, the financial and the cultural, the Native and the non-Native. As Carl Marrs (2002), president of the ANCSA CEO Association, put it, "Native people are carving their own cultures into ANCSA … [and] have begun a process of developing enduring institutions that provide important services to Alaska Natives." At the same time, the culture of corporate capitalism has put its own stamp on Alaska Natives, who are now not only hunters and gatherers but also CEOs and shareholders, often finding these roles difficult to balance. This balancing act is reflected in ANCSA corporation mission and philosophy statements, as the following examples (excerpted from their Web-sites) illustrate:

- Arctic Slope Regional Corporation (Inupiaq Eskimo, Northwest Alaska): "ASRC is committed to preserving the Inupiat culture and traditions which strengthen both our shareholders and ASRC. By adhering to the traditional values of protecting the land, the environment and the culture of the Inupiat, ASRC has successfully adapted and prospered in an ever changing economic climate" (n.d.).
- Sealaska Corporation (Tlingit & Haida regional corporation, southeast Alaska): "Sealaska's corporate philosophy is to protect and grow these assets and to use them to provide economic, cultural and social benefits to current and future Sealaska shareholders and their descendants" (n.d.).
- Sitnasuak Corporation (Inupiaq Eskimo regional corporation, Northwest Alaska): "Sitnasuak's corporate values are deeply rooted in the traditional values and ethical beliefs of the Inupiaq culture. These are the ultimate source of our corporation's strength and power. With guidance and support from our Elders, we seek to always live out these values: hard work, cooperation, sharing, respect for others, responsibility to the 'Tribe', avoidance of conflict, patience, humility, spirituality, reverence toward nature, humor, commitment to the family, love of children, respect for elders, open communication … Our tradition of cultural diversity, coupled with a strong core of traditional Inupiaq culture, had created a vitality and openness in our corporation that makes us eager to take on new challenges and pursue new opportunities. Sitnasuak's mission is to earn profits on operations while protecting our land, culture and benefiting shareholders"
- Calista (Yup'ik Eskimo regional corporation, western Alaska): "Linking our proud past to a successful future through strong corporate leadership that advocates regional unity while enhancing our cultures and lands" (n.d.).

In an age of globalization where corporations are not only increasingly mobile but also deterritorialized through processes of flexible accumulation (Harvey 1989), ANCSA corporations' emphases on multistranded, supraeconomic ties to particular peoples and places seem out of tune. On the other hand, for those who see the modern global corporations in a negative light, as powerful "alien entit[ies] with one goal: to reproduce money and replicate [themselves]," thus pitting themselves "against people–in terms of reduced job security, wage roll-backs and environmental destruction" (Korten 1995), these Native American corporations may offer a viable alternative. Ideologically, at least, ANCSA corporations posit themselves as rooted institutions that will empower Native people, nourish local culture, and conserve places.

Evaluating ANCSA Results in the First Thirty Years

Idealizing a corporation as uniquely rooted is one thing, but realizing that vision at the practical level has proved a challenge. How have ANCSA corporations managed? The results are definitely mixed. Following a recent report by Marrs (2002), we can evaluate ANCSA corporations according to conventional business criteria, such as employment, investment, and income. At the same time, we can employ nonconventional criteria, such as identity, cultural, and social capital yields, to assess the performance of ANCSA corporations in broader terms. First let us examine the conventional benchmarks.

Investment and Income

With enthusiasm many corporations initially sought to invest in developing their own lands, but were advised (often by outside business consultants) to pursue traditional forms of natural resource extraction, for example, timber, mining, fish processing, and oil, at nonsustainable levels for short-term profit. This, in turn, created big windfalls and demands for large dividends, which also could not be sustained. Other corporations, especially small remote village corporations, often had minimal resources to develop and little experience with investing money. Others were delayed in gaining land conveyances and thus lacked capital to meet even basic expenses. In one case a successful Native shareholder lent his village corporation more than $100,000 when it could not obtain financing through conventional channels due to a lack of assets. In general, small village corporations did not fare well, and in 1976 Congress amended ANCSA (acknowledging their mistake in pushing all villages to form their own corporations) to allow corporations to merge with each other. Fifty-two mergers and consolidations took place over the next five years, many to avoid or reorganize from bankruptcy.

Some village corporations deliberately chose lands away from their traditional territories so as to avoid conflicts with subsistence and other non-commercial land uses. This created conflicts between corporations. For example, the Southeast Tlingit village of Angoon, together with the Sierra Club, sued the Sitka village corporation, Shee Atiká, for their attempt to clear-cut timber on 22,422 acres of land they had selected on Admiralty Island within what Angoon, which is home to another ANCSA corporation, Kootznoowoo, considered its traditional territory. Although Shee Atiká eventually prevailed, the litigation took nearly a decade to resolve, and the corporation's costs in defending the suit exceeded $10 million (Kitka 1999; see also Metcalfe 2000). Critics cite such cases as evidence that ANCSA corporations, rather than uniting and empowering Natives, have too often divided and pitted them against each other.

In the first twenty years of ANCSA, the twelve regional corporations lost nearly 80 percent of their original cash endowment, almost $400 million (Colt 2001). In 2001, thirty years after ANCSA, eight of the twelve land-holding regional corporations had a combined net income of $167 million. But the other four sustained net losses of $134 million, making the total ANCSA balance sheet for regional corporations considerably less impressive (Marrs 2002).

Significantly, proportionally less ANCSA capital is invested in the state today than twenty years ago. This is due in part to lower returns and higher rates of failure among local investments, as well as the fickle, boom and bust character of Alaska's natural resource economy. Gary Moore (1997), a former director of the nonprofit Tanana Chiefs Conference, notes:

> Unfortunately, America's corporate system misleads Native for-profits to pursue a mission of generating revenue for the shareholders, which does little to directly benefit villages or their local economies. The resulting effect of this inappropriate system is that the Native corporations tend to shy away from investing within their own village economies or region. Any investment or development of Native lands, is without a doubt, high risk because of remote geographics alone, not to mention numerous other obstacles which inhibit profitability.

Still, despite bankruptcies, mergers, and "Not In My Backyard" (NIMBY) land selections, ANCSA corporations to date have managed to hold onto the bulk of their 44-million-acre land base. This resilience is undoubtedly linked to the ideological tenet that land is basic to Alaska Native identity, culture, and well-being and is thus inalienable.

Employment

Most ANCSA corporations seek to create employment opportunities for shareholders, and some have a shareholder preference in hiring. But to date these priorities have not translated into high rural employment. Most jobs are in the urban and regional centers, not in the villages, where half of Alaska's 80,000 Natives still reside. While sixteen of the ANCSA corporations are among the top hundred employers in the state, the other 165+ corporations are not, and many employ few or no full-time workers in rural areas. In general, the assumption that Native corporations have sacrificed profits to protect or create jobs is not supported by the data (Colt 2001). Similarly, jobs created by ANCSA corporations have not necessarily translated into Native hiring. A recent survey of 23 top ANCSA corporations (12 regional and 11 village) found that of their 10,600 employees, only 3,400 (about 30 percent) are Alaska Natives, a disappointingly low number (Marrs 2002).

Cultural Projects

Among the biggest ANCSA successes are cultural projects. Nearly every corporation has a self-imposed mandate to invest in its youth and cultural heritage. Many ANCSA regional and village corporations have developed nonprofit heritage foundations to support cultural survival. These organizations raise millions of dollars in scholarship money for youth, run Native language institutes, and publish literary and scholarly works on their culture. In addition, they organize, support, and facilitate major cultural events. For example, Sealaska Heritage Institute in Juneau, a nonprofit arm of Sealaska Corporation, runs numerous Tlingit and Haida language institutes each year, sponsors the publication of bilingual language and cultural materials, and underwrites major cultural events. Among these events is a lavish biennial festival known as Celebration, which plays host to thousands of dancers from Alaska and beyond and draws tens of thousands of spectators from all over the Pacific Northwest to the state capital of Juneau.

An important role that corporations and their nonprofit arms perform is to identify and manage cultural resources on ANCSA lands. In the early days of ANCSA, corporations worked prodigiously to identify thousands of historic sites on and off ANCSA lands. In cases where historic sites could not be selected through ANCSA, corporations have often sought other ways to protect them. A case in point is Auke Cape, a historic Tlingit site in Juneau slated for development that the author (Thornton 1997) recently evaluated for protection as a Traditional Cultural Property (TCP, a federal designation recognizing the significance and eligibility of cultural landscapes for the National Register of Historic Places). The significance and eligibility of the site having been affirmed, Sealaska Heritage Institute has taken the lead in assisting the local Tlingit tribe in pursuing the nomination of the landscape for federal and state recognition as a TCP under provisions of the National Historic Preservation Act.

Another important cultural project receiving corporate assistance is the return of human remains and objects of cultural patrimony currently residing in museums and eligible for repatriation under the terms of the federal Native American Graves Protection and Repatriation Act (NAGPRA) of 1990. As with the funding and implementation of language and heritage programs and the identification and maintenance of key cultural landscapes, repatriation efforts require significant fiscal and human resources, as well as experience in dealing with state and federal bureaucracies, in order to be successful. Working with local tribes and clans, corporations have been successful in returning and helping care for sacred objects and human remains that had been alienated, often without authorization by grave robbers and unscrupulous collectors, and housed in faraway museums.

A priority for regional nonprofit organizations that has developed in the post-ANCSA era is that of health. Most of these entities chart a holistic approach to the maintenance of Native health and well-being. As the Copper River Native Association (n.d.) puts it in their mission statement:

> Our purpose is to represent our tribes and communities by being united, responsible, sober leaders by protecting a healthy environment, which provides good health, spiritual and mental wellness, and to provide education, employment, training and to have cultural heritage education.

> To promote, protect and provide sovereignty, hunting, fishing and all other Indian Rights through representation of Native people and positive role models utilizing the wisdom of the elders and our traditional chiefs to lead the Native people to unity.

Maintaining this kind of holistic vision may be the most important project of all for corporations and their nonprofit counterparts. As Margie Brown (2001), a Yup'ik Eskimo and member of the board of directors of the Cook Inlet Regional Corporation, observes: "The regional non-profits are becoming very significant entities in Alaska Natives' lives, producing tremendous benefits and providing services. I think that will only get stronger. I see a whole series of organizations designed to address each area's specific needs."

Political Clout

Perhaps most importantly, ANCSA has made Alaska Natives a political force to be reckoned with. Despite increasingly minority status (currently 15 percent of the population and falling), Alaska Natives' strong socioeconomic organization as tribes and corporations has given them significant political influence. As Brown (2001) asserts:

> The ANCSA structure and the entities that have spun out after ANCSA have given Alaska Natives a powerful economic and political voice to speak and be heard. It took many years before there was a real acceptance in the Alaska community—many, many years before you saw a story in the local Anchorage paper about the positive benefits of ANCSA and what Alaska Natives and Alaska Native corporations were doing. That was slow in coming, but I don't think anybody doubts there are strong

economic and political entities that can now speak for Alaska Natives and they're heard in a way I don't believe they ever would have been without ANCSA.

Initially most corporate Indian leaders were politically conservative, often siding with dominant business and political interests. In 1995, for example, the corporate-dominated Alaska Federation of Natives (AFN), a political advocacy organization, openly supported oil drilling in the Alaska National Wildlife Refuge, despite vociferous opposition from Native villages and tribes dependent on the caribou that calve there. Recently, more corporations have become outspoken in their concerns for health and welfare of their shareholders beyond dividends, especially in rural Alaska. For example, one CEO went so far as to advocate civil disobedience as a means of protesting Alaska's failure to forge a viable subsistence policy for Alaska Natives (Thornton 2002).

Values and Ethics

Perhaps most difficult to balance are key economic assumptions, values, and ethics that seem to underlie capitalist corporate logic (see Korten 1995), some of which are, at least on the surface, antithetical to Alaska Native cultural ideals. First, there is the assumption that humans are motivated primarily by a self-interested quest for personal financial gain. In comparatively egalitarian hunting and gathering societies, where generalized reciprocity is the norm, the quest and accumulation of personal wealth can earn one considerable enmity and sanctions. The Arctic Slope Regional Corporation, among the largest and wealthiest in the state, says it instills the traditional Inupiat values in its subsidiaries, employees, and shareholders, which it considers part of the Inupiat "extended family." Among these values is "Honoring the philosophies of cooperation and sharing." Elsewhere, the corporation declares: "We still respect the environment" because "This is our home." This ethos of reciprocity and respect for homeland, resources, and people, which permeates all Alaska Native worldviews, remains foundational to the development of sustainable communities and environmental ethics, despite corporate involvement in industrial development. And where industrial development appears to violate this ethos, as in the case of the recent state push, with little local input, to accelerate oil drilling in fragile coastal subsistence areas, dissenting voices can be heard within the Native community, including that of the Inupiaq mayor of the North Slope Borough, George Ahmaogak, who said: "Lately, we've been feeling like our seat at the [management] table is having one of its legs cut off" (Dobbyn 2004b).

Second is that idea that actions that yield the greatest financial gain for the individual are also the ones most beneficial to society. There is still a strong distrust of money accumulation and individual financial gain; individuals are judged by the communal benefits they render through traditional (hunter, clan leader, etc.) and modern socioeconomic roles. This is one reason why numerous ANCSA corporations have suffered backlashes from dissident shareholders. These shareholders, often the poor and marginalized, attack the status, wealth, power, and perks that have accrued to corporate executives without comparable benefits to shareholders. A recent petition by dissident shareholders of Kootznoowoo Village Corporation in the low-income village of Angoon sought a special dividend of $15,000 (Dobbyn 2004a). Leaders of the movement claimed that corporate executives had squandered ANCSA monies on failed business ventures while the shareholders reaped little or no reward. They wanted remaining assets to be liquidated and the $9.4 million in cash paid out to individual shareholders to invest on their own. Despite two hundred signatures, the petition did not succeed, although it did raise a very public debate about (perhaps inflated) shareholder expectations and how well ANCSA corporations have been serving communities.

Third is the belief that competitive behavior is more rational for the individual and the group than cooperative behavior. Cooperative behavior is considered the hallmark of Alaska Native kin-based socioeconomic and political structures, even among traditionally stratified Northwest Coast cultures. There is a strong emphasis on individual competence and autonomy but not on competitiveness. This concept is embodied even in the very names of Native corporations. Calista, for example, the name of the southwest regional ANCSA corporation, is a Yup'ik Eskimo term that may be translated as a collection of people (*ista*) "working" (*cali*) together.

Fourth is the idea that human progress is best measured by members' incomes and the value of what they consume. Many elders look to the health of their loved ones and their lands as the best measure of success. Continuity with the past and social and cultural capital are stressed as much as material progress, individual wealth, and consumption. Participation in the subsistence economy, for example, is widely viewed as the basis not only for conserving natural resources and relationships to homeland but also for maintaining social capital and civic virtue (Putnam 2000: 19). As Tlingit elder and former ANCSA corporation vice president Herman Kitka (1998: 48) puts it:

> Teaching subsistence occurred where our customary, traditional subsistence food supply was collected. *This education was what made each Tlingit a good citizen in each community.* The young people learn to respect the land they live on. They also learn to take only what each family needs to make it through the year. We need to keep teaching the children our subsistence lifestyle, culture, and religion. Without this education, our Tlingit cultures will be lost forever. [emphasis added]

Negotiating the balance between mainstream capitalist corporate values and Native communal values requires both ideological and practical work. There are potential but not insurmountable contradictions. The potential contradictions lie partly in the imagination, in the ways we tend to construct diametrically opposed images of the traditional, pure, and noble subsistence-oriented indigene versus the modern, sullied, and alienated corporate Indian no longer tied to his land. Natives and non-Natives alike, from dissident shareholders to ANCSA corporation personnel to environmentalists and legislators, play upon and manipulate these images to their own ends. But the images are inaccurate, static one-dimensional stereotypes; in truth the process of negotiation and exchange between Native subsistence and corporate economies, cultures, and identities is leading to new understandings about what constitutes a sustainable community and what it means to be an Alaska Native. This is not a new process, either. As anthropologist Eric Wolf (1982: 17-19) famously observed:

> What of the localized Algonkin-speaking patrilineages ... which in the course of the fur trade moved into large nonkin villages and became known as the ethnographic Ojibwa? What of the Chipeweyans, some of whose bands gave up hunting to become fur trappers, or "carriers," while others continued to hunt for game as "caribou eaters," with people continually changing from caribou eating to carrying and back?... The tacit anthropological supposition that people like these are people without history amounts to erasure of 500 years of confrontation, killing, resurrection, and accommodation.... The more ethnohistory we know, the more clearly "their" history and "our" history emerge as part of the same history.

For this reason, no one can realistically advocate for a return to some idealized, precontact form of subsistence, just as no one can realistically expect ANCSA corporations to behave exactly like non-Native corporations. Yet there is a plethora of views within Alaska Native communities and corporations on how, and perhaps even whether, to balance traditional sociocultural and modern economic forces, and this has led to conflicts and "fragmegration," or fragmentation and new forms of integration, within and between constituencies and communities. At the same time, however, the increasing urgency of resource conflicts and conservation issues seems to be stimulating greater dialogue, if not a consensus, on the need to develop more sustainable local economies based not merely on money but also on the social, cultural, and natural capital inherent in subsistence-based Alaska Native cultures. As former Sealaska Corporation CEO Mallott (2001b) concludes: "The [ANCSA] corporations must be used as tools for our future, as opposed to instruments to determine our future. They must be guided by our values and our spiritual heritage."

Conclusion

In a probing essay on the fates of hunting peoples in developed states, anthropologist Harvey Feit (1983: 404) argues:

[W]hen hunting peoples in developed states are able to mobilize some political/economic leverage in the macro-arena then it may be possible for them to resist external pressures leading to a restructuring of their own social fabric, but they may also be able to restructure the relationship between themselves and the impinging institutions. The extent of such a restructuring is variable, and the means of reorganizing relationships between hunting society and macro-societies necessarily involve the creation and introduction of new institutions.

The post-land claims settlement era in Alaska has been a grand experiment in creating such institutions. What lessons can we draw from thirty years of this ANCSA experiment?

The economic results of ANCSA corporations definitely are mixed, with some clear successes and some magnificent failures. However, the people- and place-centered focus of ANCSA corporations has yielded distinct benefits to Alaska Native peoples and to the state. Unlike conventional corporations, which often move and "globalize" their operations to reduce costs at the expense of local people and local economies, ANCSA corporations remain inextricably tied to the aboriginal lands and peoples of Alaska. At the same time, although ideologically and structurally different from conventional corporations, ANCSA corporations are subject to the same economic logic and cycles—market swings, capital flows, and so on—that affect all businesses. Thus, the economic vulnerability of Alaska Natives may be exacerbated rather than buffered by ANCSA corporations subject to global market conditions.

The subsistence economy, meanwhile, has maintained itself in rural areas and enjoys a growing ideological and political support among Natives, who are unfortunately an increasingly small minority of Alaska residents, and even powerful ANCSA corporate executives. As Tlingit Indian and former CEO of Sealaska Corporation Robert Loescher (1999) stresses, a viable subsistence policy for Natives is imperative and something that all Native entities, including ANCSA corporations, must defend:

> Is [subsistence] … worth fighting for? The answer is yes. Maybe more so than any fight or challenge ever faced by us as Alaska Native people. … But we must face this challenge with the willingness to put it all on the line. There is no longer any room in this fight for political in-fighting, back room deal-making, concessions made for the sake of business, or for those who would divide rather than unite. Now is the time to put to use all that we have garnered and developed over the years: our political sophistication, our economic stature, our collective will. They must all come to bear if we are to prevail. Will ours be the generation to preserve or let disintegrate the subsistence legacy?

The subsistence legacy, as framed by Loescher, is the legacy of ecologically sustainable, life-enriching local economies—healthy communities based on renewable wild resources and limited commercial development.

A similar, broad vision of healthy communities and corporations is put forth by Loescher's predecessor at Sealaska, Byron Mallott (2001b):

> As we look to the future and search for the path of our journey, we ourselves need to sit down and say, "Who are we? Where are we going? What do we want?" Let's look to the year 2050 with a vision of healthy communities, a rural Alaska that is alive and vibrant. … Our children have to have access to quality education that is responsive socially and culturally … our vision must be that we live with self-esteem and pride because Alaska society and its social policy respects who we are. In the future, we must be recognized not as separate, but as being different and therefore making the whole of Alaska richer by our being here.

Greater recognition among corporate and governmental leaders of the importance of Alaska's environmental, cultural, and economic diversity and Natives' spiritual heritage in the land is a key step toward "rerevolutionizing" the idea of the business corporation as a rooted, communal institution, and toward reproducing healthy, sustainable communities in rural Alaska. Through this revaluation process the seemingly paradoxical forces of subsistence and corporate economies might be reconciled in ways that could serve as a model for sustainability in communities beyond Alaska facing similar challenges in blending traditional values with modern imperatives to sustain local networks and cultural lifeways amid global economic development and social change.

Note

1. Some Alaskan cities, such as Juneau, were founded on aboriginal Native villages and became largely non-Native urban centers as a result of subsequent development. In such cases, local Natives became ineligible for subsistence under the rural preference.

References

ADF&G. 2004. Alaska Department of Fish & Game, Division of Subsistence. Community Profile Database. Juneau: www.subsistence.adfg.state.ak.us/ geninfo/publctns/cpdb.cfm.

Arctic Slope Regional Corporation. (n.d.). Retrieved May 17, 2004, from www.asrc.com/ home/home.asp.

Berger, Thomas. 1985. *Village Journey*. New York: Hill & Wang.

___. 1991. *A Long and Terrible Shadow: White Values, Native Rights in the Americas since 1492*. Vancouver/Toronto: Douglas & MacIntyre.

Brown, Margie. 2001. Interview with Sharon McConnell. .

Calista Corporation. (n.d.). "Cultural Heritage." Retrieved May. 17, 2004, from http://www.calistacorp.com/heritage.html.

Colt, Steve. 2001. "Alaskan Natives and the 'New Harpoon': Economic Performance of the ANCSA Regional Corporations." Anchorage: Institute of Social and Economic Research.

Copper River Native Association. (n.d.). Retrieved May 17, 2004, from
http://www.copperriverna.org.

Dobbyn, Paula. 2004a. "Corporation Stockholders Want Payout." *Anchorage Daily News*,
March 24, p. F1.

___. 2004b. "Development Crowding Slope, Mayor Says." *Anchorage Daily News*, February
12, p. F1.

Dombrowski, Kirk. 2002. "The Praxis of Indigenism and Alaska Native Timber Politics."
American Anthropologist 104 (4): 1062-73.

Feit, Harvey A. 1983. "Conflict Arenas in the Management of Renewable Resources in the
Canadian North: Perspectives Based on Conflicts and Responses in Northern Quebec."
Paper presented at the Third National Workshop on People, Resources, and the
Environment North of 60o. Yellowknife, June. Ottawa: Canadian Arctic Resources
Committee.

Goldschmidt, Walter R., and Theodore H. Haas. 1998. *Haa Aaní, Our Land: Tlingit and
Haida Land Rights and Use*, ed. Thomas F. Thornton. Juneau and Seattle: Sealaska
Heritage Foundation and University of Washington Press.

Harvey, David. 1989. *The Condition of Postmodernity: An Enquiry into the Origins of
Cultural Change*. Oxford: Blackwell.

Hensley, Willie. 2001. Interview with Sharon McConnell. University of Alaska Anchorage.

Kitka, Herman. 1998. "Deep Ties to Deep Bay: A Tlingit Elder's Training." *Cultural
Survival Quarterly* 22 (3): 47-48.

___. 1999. Interview with author. Sitka, Alaska.

Korten, David. 1995. *When Corporations Rule the World*. Hartford, CT: Kumarian Press.

Loescher, Robert W. 1999. "Native Subsistence Rights—Where Are We Now in State and
National Politics?" Speech/paper presented to the Alaska Federation of Natives Political
Leadership Summit, February 16, 1999. Juneau: Sealaska Corporation.

Mallott, Byron. 2001a. Interview with Sharon McConnell. University of Alaska Anchorage.

___. 2001b. "Unfinished Business: The Alaska Native Claims Settlement Act." University
of Alaska Anchorage. http://litsite.alaska.edu/uaa/aktraditions/business.html.

Marrs, Carl. 2002. "Native Corporations: An Epic Story Benefiting Alaska." *Commonweal
North Forum*, March 21, 2002. Proceedings. .

Metcalfe, Peter. 2000. *Earning a Place in History: Shee Atiká, the Sitka Native Claims
Corporation*. Sitka: Shee Atiká Corporation.

Mitchell, Donald Craig. 2001. *Take My Land, Take My Life*. Fairbanks: University of Alaska
Press.

Moore, Gary A. 1997. "Leaders Need to Acknowledge Complexities Created by ANCSA."
Fairbanks: Tanana Chiefs Conference. Retrieved May 17, 2004, from
http://arcticcircle.uconn.edu/SEEJ/Landclaims/moore.html.

Putnam, Robert. 2000. *Bowling Alone: The Collapse and Revival of American Community*.
New York: Simon & Schuster.

Sealaska Corporation. (n.d.). "About Us." Retrieved May 17, 2004, from
http://www.sealaska.com/aboutus.htm.

Sitnasuak Native Corporation. (n.d.). "About Us." Retrieved May 17, 2004, from
http://www.snc.org/aboutus.htm.

Thornton, Thomas F. 1997. *Traditional Cultural Property Investigation for Auke Cape, Alaska.* Report filed with the National Oceanic and Atmospheric Administration (Project No. 601.00, Contract No. 50ABNA600056), and in author's possession.

___. 1998. "Alaska Native Subsistence: A Matter of Cultural Survival." *Cultural Survival Quarterly* 22 (3): 29-34.

___. 1999. "Subsistence: The Politics of a Cultural Dilemma." In *Public Policy Issues in Alaska: Background and Perspectives,* ed. C. Thomas. Juneau: The Denali Press.

___. 2001. "Subsistence in Northern Communities: Lessons from Alaska." *The Northern Review* 23: 82-102.

___. 2002. "From Clan to Kwaan to Corporation: The Evolution of Tlingit Political Organization." *Wicazo Sa Review* 17 (3): 167-94.

Wolf, Eric R. 1982. *Europe and the People without History.* Berkeley: University of California Press.

Wolfe, Robert J. 1998. "Subsistence Economies in Rural Alaska." *Cultural Survival Quarterly* 22 (3): 49-51.

Wolfe, Robert J., and Linda Ellanna (compilers). 1983. "Resource Use and Socioeconomic Systems: Case Studies of Fishing and Hunting in Alaskan Communities." *Alaska Department of Fish and Game, Division of Subsistence Technical Paper Series,* No. 61. Juneau.

Communities Out of Place

Johanna Gibson

Introduction

This land is mine
This land is me

These two lines come from the Australian musical *One Night the Moon*, directed and coscripted by Australian indigenous filmmaker Rachel Perkins. Based on actual events in the Australian outback in 1932, the film tells the story of a pastoralist couple whose child, after waking up at the bright moonlight streaming into her bedroom, wanders into the night chasing after the moon and is lost in the bush. The father refuses the help of an Aboriginal tracker and orders him off his land, an action with tragic consequences. In this scene, the father's refrain, "This land is mine," is juxtaposed by Albert the Tracker's refrain, "This land is me."

This scene and this refrain capture one of the key differences between conventional proprietary models and traditional custodianship. The fundamental concern for devising appropriate recognition and protection for indigenous cultural resources in the land is the imperative toward reconciling that protection within already existing forms of ownership under conventional laws of property and intellectual property (*this land is mine*). However, solutions based on consistency with such laws, for the purposes of protecting cultural and traditional knowledge, are contrary to customary laws of indigenous Australian custodianship with respect to land and the traditional and cultural knowledge found within that land (*this land is me*). According to customary Aboriginal laws, as explained by a senior traditional lawman of the Ngarinyin people, "I don't own the land, but the land owns me ... That's why it's so important for us, because the land owns us" (Mowaljarlai 1995).

While conventional models of ownership vest, in the individual, rights with respect to a recognizable and exhaustible entity, diverse indigenous concepts of custodianship resist the universality of this regime and suggest various relationships of shared and enduring interaction with the land that transcend each individual and indeed the "boundary" of the parcel itself; custodianship is linked to the land but not necessarily to place, in the territorial, monopolistic sense. Real, yet intangible, nonexhaustible, and inalienable, the various forms of indigenous custodianship cannot be translated as the finite and temporary *this land is mine,* in which the enduring community in *this land is me* will expire. It is this fundamental distinction, presented so elegantly in this film, from which this discussion of communal rights will proceed. The Australian example, drawn upon in this essay, is particularly important in presenting the tension between, on the one hand, indigenous, communal values, and on the other hand, the trading sovereignty of a developed country and the anxiety of national dominion over natural resources.

Real Intellectual

The traditional knowledge[1] of indigenous Australians is almost inextricably linked to the land, and yet Australian native title law has resisted the acknowledgment of rights to cultural knowledge embedded in the land. Instead, the possible recognition and protection of such knowledge remain questions referred, somewhat problematically, to applicable intellectual property laws. Similarly, the international community also resists the link between cultural knowledge and land. Attempts to resolve questions of international protection continue to interpret and construct questions of traditional knowledge primarily within the parameters of international intellectual property law, as indeed concerns of intellectual property creation and trade. The task of devising an international system of protection has been assigned to the World Intellectual Property Organization (WIPO) Intergovernmental Committee on Intellectual Property and Genetic Resources, Traditional Knowledge and Folklore (IGC).[2] The work of this committee itself suggests an ongoing cooperation with the construction of traditional knowledge protection as an issue of intellectual property law.

Arguably the "problem" is interpreted within such a forum because the major impact of any such protection will be upon trade in intellectual property and commercial access to resources. In other words, the relationship to intellectual property is not necessarily sustained because of any inherent comparability between traditional knowledge and the available categories of intellectual property protection. In fact, work toward possible solutions continues to examine ways in which to make traditional knowledge compatible with intellectual property protection, and the solutions themselves are being considered for consistency with international intellectual property laws.[3] This emphasis is maintained despite the conflict over access to land, particularly apparent in the case of Australian native title law, frequently being a question with respect to cultural resources (and knowledge) in that land.

Whether an intellectual property framework is the appropriate basis upon which to build effective protection regimes, however, is open to criticism, in that the special kinds of interests and objectives that must be met by that protection do not necessarily fall within the conventional protection regimes of intellectual property law. Organizing the protection of traditional knowledge around the principles of intellectual property law necessarily results in the delimiting of communal or "public"[4] knowledge within the conventional private ownership that is protected by intellectual property regimes. In other words, it forces the cultural values of identity and community cohesion (*this land is me*) within the paradigm of individual ownership (*this land is mine*).

An intellectual property framework cannot begin to realize the interaction with (and response to) resources that is arguably constitutive of traditional and indigenous community, very broadly speaking, rendering the expropriation of resources an issue not of sustainability and management of resource-objects but a threat to the existence of the traditional and indigenous communities themselves. The exchange of knowledge and resources, therefore, is not merely an issue of international trade in knowledge and property but involves traditional access and use of land, the right to self-development of traditional communities, and finally the relevant and appropriate protection of the cultural and intellectual value in resources and their use.

The imposition of the intellectual property model extends even to the imposition of categories upon the species of traditional and indigenous resources, categories that are doubtful in their relevance to indigenous groups themselves.[5] This includes the often-inappropriate distinction imposed by an intellectual property framework itself, where traditional knowledge approximates industrial property in its usefulness or commercial value, and folklore suggests creative works for which copyright (and the attending "celebrity" of the author) may be the appropriate protection.

This distinction itself is related to the different forms of ownership and exclusive rights that attend those different properties, and indeed the different "values" calculated for those properties, and realized in the form of different periods of monopoly as defined by intellectual property laws. Within intellectual property regimes, the owner is assuming a key position. Arguably it is not until the litigation and enforcement (or exercise) of intellectual property rights that those rights and that property assume any real effect. Thus, intellectual property regimes are intrinsically reliant upon the model of private ownership and the normative principle of individual rights to exercise the benefits of that ownership, and indeed to realize the "property" in that intellectual activity.

Pursuing the distinctions imposed by the intellectual property models, and presuming the applicability of Western objectives to indigenous groups, will continue to override and overlook the way in which biological and technological resources are related to the cultural expression that facilitates the cohesion and integrity of a particular group. Furthermore, constraining the quality of indigenous and traditional knowledge within objects capable of "commoditization," according to international intellectual property laws, is potentially disparaging and arguably damaging to the crucial interaction between community and resources.

Considering the contrast between private ownership rights of intellectual property and the communal obligations of custodianship that must be given legal effect, the present discussion will demonstrate the possible irrelevance to communal custodianship of a strictly intellectual property model of protection. It will be suggested that a community model of protection must be developed beyond the possessory relationship protected by intellectual property regimes.[6] An emphasis on production and selfhood that persists in the celebrity of the individual in exclusory private rights, as distinct from the obligations that circulate between individuals in an inclusive dialogue of community, will merely sustain the normative principle of the individual that informs the intellectual property model. This model will never be able to account for the diversity of meaning in "ownership" or "custodianship" that must be enlivened if fair and equitable recognition and protection is to be achieved.

Communal Inclusion, Private Exclusion

The possessive relationship that justifies and sustains intellectual property law exists in a reciprocal legitimation with the individualistic sense of self that dominates Western legal paradigms (Underkuffler 2003). This conundrum within conventional legal paradigms operates between the normative individual's possessive relationship to property and the generation (through the very exercise of that individual production and possession) of the individual legal subjectivity that is necessary to access such rights. This raises the significant obstacle of how to conceive of community "selfhood" outside such models of possession but nevertheless ensure a particular community's entitlement to protective mechanisms within international legal models.

Conceiving of community in terms of collective ownership and property rights (within an intellectual property model) risks rationalizing that community within a market economy and assimilating its interests and values within a discourse of individual private rights. A collective, for the purposes of conventional proprietary models of creativity and innovation, is a group within the same geohistorical location, existing together and creating together at the same time and in the same place.[7] Similarly, it is suggested elsewhere that to rely upon the collective rights, arguably recognized under international human rights law also fails to recognize this limitation and ultimate individualization of the "self" of that collective (Charlesworth & Chinkin 2000; Donnelly 1989), for the purposes of traditional knowledge protection and the values of community.

Indeed, this assimilation of community within intellectual property models is made possible through what might be considered an oversimplification of the communal forms of resource and knowledge management within indigenous and traditional communities. Most discussions of traditional knowledge within the intellectual property paradigm assume that ownership in traditional communities is communal (see Dutfield 2004: 95-96). By making this assumption, such discussions almost invariably arrive at a problematic communal/individual polarization of what is nevertheless a conventional model of property ownership. While communal models are often appropriate, it is more accurate to be aware that custodianship of particular knowledge is differentiated within a particular community according to communal values but not necessarily entirely open to access by all members of that community. Thus, traditional knowledge is not necessarily held by all members within a particular community, and is not necessarily accessible by all members of a particular community, but rather is held according to the differentiation within that community:

> Although individual creativity is not stressed in individual communities, it would be wrong to jump to the extreme and suppose that designs are subject to a generalized communal right. Communities are internally differentiated to quite a high degree, and their members should not be seen as interchangeable units. On any matter, some people are likely to have rights of a certain kind, others rights of another kind, and yet others no rights at all. (Maddock 1988: 9)[8]

However, that knowledge is held by one or by several in accordance with the values of the community:

> The impetus for the creation of works remains their importance in ceremony, and the creation of artworks is an important step in the preservation of important traditional customs. It is an activity that occupies the normal part of the day-to-day activities of the members of my tribe and represents an important part of the cultural continuity of the tribe.[9]

In other words, ownership may or may not be communal, but the regulation of that ownership is communal in the strictest sense; that is, management of resources (including those in knowledge) will be according to customary law and traditional values shared by the community.[10] On the other hand, the regulation of intellectual property models is undertaken externally. Thus, what is of primary importance to the development of adequate *sui generis* rights of community is to acknowledge that genuine communal management is significant.

Therefore, a system of relevant and effective protection of traditional knowledge may require the recognition of customary law by intellectual property systems, rather than incorporating *sui generis* elements of an intellectual property system in order to make "customary" laws consistent with that regime. In other words, rather than assimilating traditional knowledge within intellectual property, to make its protection consistent with those regimes (*this land is mine*), appropriate international protection arguably must consider the ways in which intellectual property law can be made consistent with customary laws pertaining to traditional resources (*this land is me*).[11]

Therefore, an appropriate model may be one that vests the authority for management and regulation of ownership of traditional knowledge in the community according to customary laws. Fundamentally, conventional Western legal systems must trust customary law rather than disregard or presume customary law to be disorganized, fluid, and unenforceable. It is necessary for the international trading community of states to respect the effectiveness and certainty of indigenous and traditional individuals' obligations to community and allegiance to their culture such that customary law will bind individuals with the requisite legitimacy within any international system.

Understanding the relationship between cultural expression and the land is critical to this development of the concept of community and to concretizing the responsibility to cultural diversity through its interaction with biological diversity. This development in the concept of community facilitates not only access to rights in indigenous and traditional resources but also cultural diversity as well as capacity building at the level of community. Furthermore, in recognizing the link between traditional cultural practice and community cohesion, and the natural resources of that community, biological diversity is rendered coincident not merely with the geophysical area or, for that matter, with the artificial boundaries of nation-states but rather with the topological development of different and overlapping social spaces of community. Thus, communities and their social and cultural development are essential to approximating a complete picture of global biodiversity because that biodiversity is a social and political knowledge and not merely a natural resource over which nation-states may exercise permanent sovereignty.

The major instrument arguably linking cultural and biological diversity, and through which such international legitimacy of indigenous and traditional communities might be achieved with respect to biological diversity, is the Convention on Biological Diversity (CBD) administered by the United Nations Environment Program (UNEP). Despite the acknowledgment of indigenous and traditional knowledge in this significant international agreement, it is important to be aware of the limitations of principles established in the CBD. In particular, the preamble of the CBD asserts that states have sovereign rights over their biological resources, as distinct from those resources being rendered global public goods. Despite the importance of this provision to developing countries, in the use of access and benefit-sharing agreements between countries, it is nevertheless highly problematic for traditional and indigenous communities. The autonomy of indigenous and traditional communities is subject to territorial sovereignty over natural resources, rather than being realized through the recognition of the rights of the community in its resources within an international framework. Thus, community is made subject to trading relations between states, instead of being respected through recognition of customary rights in resources prior to and beyond the imposition of the interests of contracting nations.

Providing effective legal and political opportunity for indigenous and traditional communities to pursue cultural and customary laws and expression in a contemporary sociopolitical context, rather than rationalizing those resources within the boundaries and identity of nation-states, is essential not only for community integrity but also for fulfilling responsibilities and obligations to cultural and biological diversity. Beyond the dominant international context of intellectual property, how might international law create a framework to which the protection and management of traditional knowledge may be referred with certainty and with cultural relevance and specificity? How might community be realized within an international system?

Creating the Potential for Property

A community model will not be effective if it persists as a nostalgic and moralizing "protection" of traditional community. In this sense, it would merely represent the geohistorical fixation of a "collective" and could never realize the potential of communal custodianship nor recognize the community in a contemporary context of ongoing cultural and intellectual expression. A community model must offer a development of the concept that will have authority and capacity within a contemporary legal framework; but at the same time, the community model will shift the emphasis from that of localized and exclusory individual property of the creator (or owner) to that of community relationships beyond place and property.

Recognition of community relationships will be respectful of the continuing evolution and contemporary identity of a particular indigenous or traditional group in its interaction and management of its resources (Golvan 1992:5). The autonomy of community must be located in the interactions rather than deferred by the physical property posited as necessary for its realization. In effect it is this "personality" of the community, which is contained in the relations between individuals rather than in the historical identity of the group that must be given legal effect as a subject, or indeed personality, within an international framework for protection.

Owning the Realized Rights to Property

As introduced earlier, it is inaccurate to argue that there is no "ownership" within indigenous or traditional communities, although the systems of management may not be compatible with the exclusory competitive systems constructed in intellectual property frameworks:

> While IPRs [Intellectual Property Rights] confer private rights of ownership, in customary discourse to "own" does not necessarily or only mean "ownership" in the Western non-Indigenous sense. It can convey a sense of stewardship or responsibility for the traditional culture, rather than the right merely to exclude others from certain uses of expressions of the traditional culture, which is more akin to the nature of many IP [intellectual Property] rights systems.[12]

To continue to frame decisions on indigenous and traditional knowledge (including cultural knowledge in the land) within this assumption against "ownership" is to deny custodianship and legitimize the exploitation of traditional knowledge as a public resource. In fact, this is not a true acknowledgment of resources as public, in that these actions simultaneously remove that knowledge to the private domain of the particular commercial interest, through the operation of patents or through works derived from traditional methods and protected by copyright, and so on. This line of "global resources" reasoning fails to understand that customary use or communal custodianship is simply a different model of possessory-like relationships to the resources. Rather than the exclusory *this land is mine,* it presents the inclusory dialogue and communal sustenance of *this land is me.*

Becoming Community: *This Land Is Me*

In order to achieve an effective legal subjectivity for the indigenous community, the community must be able to evolve beyond the fixation of the geohistorical moment of colonization. As distinct from individual relationships to a "commodified" entity (*this land is mine*), the community is realized through the ongoing and enduring relationships between members despite changes to its particular constitution over time. The identity of community, as it were, persists in the integrity of cultural and traditional expression and use of resources: *this land is me.*

The concept must be mobilized beyond the historical, physical, localized, geographical locus. In order to realize access to the public sphere, and to enjoy authority to act within that sphere, the community cannot be known, captured, translated, and incarcerated in this geohistorical locus but must be self-determined, self-recognized, and self-regulated. The problem persists, however, that "[t]he traditional sociological approach implies that when community is stripped of the local spatial dimension, it becomes such an elusive idea that it becomes virtually meaningless" (Little 2002:57). That meaning must come not from the narration and summarization of the legal system, nor from sociological models. Rather, it must come from affording the community the opportunity to defend (and assert) itself.[13]

For the purposes of a viable international legal framework, therefore, the concept of community must be developed toward the consideration of the relations between members as the subject of communal identity and rights. This perspective enables the enduring identity and autonomy of a particular indigenous community despite its evolution and adaptation in the face of ongoing colonization and the effects of dispersal and alienation: "community should not be thought of solely as the domain of the small scale and geographically local"(Little 2002: 63).

Indeed, the evolution of a community must not preclude the protection imagined within a *sui generis* system of obligations to the practice of community itself nor supersede that protection, as it were, as has been the case in recent applications of the Australian 1993 Native Title Act (NTA).

Australian Native Title Rights: *This Land Is Mine*

The Australian NTA was enacted in 1993 to administer the *sui generis* rights in land that were recognized by the High Court of Australia in *Mabo v. State of Queensland (No 2)*.[14] This decision of 1992 appeared to recognize the necessary relevance of customary law and, by extension, indigenous cultural property in the traditional use and cultural practices associated with that land such as agricultural practices, fishing sites, or the knowledge and use of medicinal plants, where that cultural knowledge (and associated practices) inheres in the land and is recognized as traditional use of the land (Puri 1993: 159). Recent decisions of the Australian High Court, however, have effectively eroded these rights and defeated the potential for these rights to protect cultural knowledge embedded in the land. The court has progressively fixed the concept of community according to its historical constitution and location, that is, as a historical commodity.

The decision of the High Court of Australia, in *Western Australia v. Ward* (2002: 313 CLR1), has been criticized as weakening significantly the strength of native title rights and their potential relevance to cultural knowledge in the land. In particular, the strict requirement of the connection to land betrays the traditional sociological notion of community invoked by the courts when applying the NTA. The application of the concept of community in this decision is of critical significance to a consideration of communal custodianship, in that it demonstrates the seriousness of limiting community to a traditional geohistorical and recognizable entity, rather than a process of cohesion and integrity between individuals through recognition and shared practices.

In *Ward*, the High Court ruled out the extension of the NTA to the protection of cultural knowledge except to the extent of controlling access to land and waters in which that cultural knowledge is situated; that is, the land as a tangible resource is rendered separate from the intangible cultural knowledge contained therein. This artificial and arbitrary separation is sustained despite the willful blindness to the interrelationships between natural resources and cultural knowledge.[15] While cultural knowledge occurs in relation to the land, the majority decision established for Australian native title law that any interests in that knowledge traveling beyond denial or control of access to land or waters will not be interests or rights protected by the NTA:

> [I]t is apparent that what is asserted goes beyond that to something approaching an incorporeal right akin to a new species of intellectual property to be recognized by the common law under para(c) of s 223(1). The "recognition" of this right would extend beyond denial or control of access to land held under native title. It would, so it appears, involve, for example, the restraint of visual or auditory reproductions of what was to be found there or took place there, or elsewhere. (p. 84)

The court held that the NTA refers not to how indigenous peoples use or occupy land or waters but to a "connection" and whether that connection can be established on the basis of the applicable traditional laws and customs. (pp. 85-86)

In this and other native title decisions, the court has relied increasingly upon the historical constitution and location of community for its recognition. This sociological and anthropological rendering of community as a historical commodity, as it were, deprives the particular community of the capacity for geohistorical adaptation and development. An effective concept of community, for the purposes of any international system of protection for traditional knowledge, must acknowledge the physical, geographical, and constitutive evolution of the particular community and recognize that its identity exists in, and is sustained by, self-recognition and communal practices. Effectively, the conventional concept of community deployed in Australian native title law ignores relationships of differentiation and locates the relevant group in the past, denying it economic and political viability in a contemporary public sphere.

The conventional concept of community betrays the kind of assimilation of community within "collective," that is, the major limitation of current models of communal or collective ownership and rights. In other words, community persists as a geohistorical moment, a particular entity–an individual. The viability of the community model is compromised by this adherence to the primacy of the normative individual entity of rights-based frameworks, whether intellectual property rights or collective rights to self-determination. Effectively, this traditional concept of community assimilates the indigenous or traditional group within a model of collective ownership as constructed by conventional property systems.

The effect of this assimilation, therefore, is particularly serious in Australian native title litigation. Community is constructed according to place, which may seem obvious given that the litigation involves *sui generis* rights to a particular place. What is not obvious is the connection requirement, where colonization almost invariably breaks the traditional community's physical connection and, in some cases, its ability to maintain the cultural connection to a place. Similarly, this connection to place cannot account for the systematic exclusion of cultural practices and uses of resources with respect to that place, through the processes of colonization.

Furthermore, traditional use by that community is constructed according to time, to the historical moment of colonization. The result is a strict definition of a group constituting that community, rather than an understanding of the internal differentiation according to values and principles that characterize the particular community. Community becomes an object of history rather than an expression of the necessary subjectivity for which entitlements to certain rights will apply. Traditional use and cultural knowledge is thereby objectified and "museumized" as an anthropological concern rather than as part of the ongoing expression of the viable community.

This position is evident in the court's finding that there were no existing interests in minerals and petroleum that could have been disturbed by the event of colonization. This was because traditional laws and customs did not extend to the modern mining of such resources at the moment of colonization. Thus, the court fixes traditional use or knowledge according to this particular historical moment, and the community is fixed as a particular historical entity in that anything it achieves or produces or expresses beyond that moment has no relevance. The community has no potential; the court has no sense of community.

In his minority judgment, Kirby J questioned this position, noting that the common law recognizes the capacity of traditional law and customs to evolve and adapt,[16] and in doing so, he suggested that the common law may incorporate the use of modern materials or resources that might have developed relevance to indigenous communities.[17] Kirby J maintained that if cultural knowledge is related to land then it must be protected for the purposes of the NTA, particularly in the context of Australia's ratification of international instruments providing for the protection of fundamental human rights,[18] including those rights to full ownership, control, and protection of cultural and intellectual property. (pp. 247-48) In addressing the relationship of the modern community to the land, Kirby J identified the artificial restriction imposed upon the community by its historicization:

> When evaluating native title rights and interests, a court should start by accepting the pressures that existed in relation to Aboriginal laws and customs to adjust and change after British sovereignty was asserted over Australia. In my opinion, it would be a

mistake to ignore the possibility of new aspects of traditional rights and interests developing as part of Aboriginal customs not envisaged, or even imagined, in the times preceding settlement. (p. 244)

Kirby J noted the majority decision that such rights are within the ambit of intellectual property laws but argued that established intellectual property regimes are "ill-equipped to provide full protection of the kind sought in this case."[19] Indeed, this minority judgment apparently supports the current discussion that intellectual property protection should adapt to accommodate traditional knowledge and customary values attached to that knowledge, as distinct from the position that insists upon protection that will be consistent with current intellectual property regimes.[20] How, then, might protection be achieved for traditional knowledge, without rendering its proposed protection merely consistent with current forms of protection for intangible property and separate from the significance of the land? How might community in the land be relevant to community in culture?

Legal Perceptions of Community

The bodies we perceive are, so to speak, cut out of the stuff of nature by our perception. (Bergson 1944)

Importantly, a *sui generis* system must resist the historical fixing in an intellectual property model, of the "identity" of the object itself and the individual ownership of that object (whether that fixing occurs through the identifiable author in traditional art, the invention in a medicinal method, the required unbroken connection to place in Australian native title, or otherwise). In the WIPO IGC, discussions toward international protection for traditional knowledge have identified this evolutionary nature of community, cultural process, and knowledge production:

[C]ultural heritage is in a permanent process of production; it is cumulative and innovative. Culture is organic in nature and in order for it to survive, growth and development are necessary–tradition thus builds the future. While it is often thought that tradition is only about imitation and reproduction, it is also about innovation and creation within the traditional framework.[21]

A workable and relevant concept of community, for the purposes of international protection, must be organic, dynamic, and evolutionary by definition, as it were. In this way, "additions" to the culture are also eligible for protection as cultural and traditional knowledge and are not limited by notions of cultural heritage and historical artifacts. As set out in the discussions of the WIPO IGC, indigenous peoples and traditional communities must "be regarded as the primary guardians and interpreters of their cultures and arts, whether created in the past, or developed by them in the future."[22] In this way, unlike the problems identified in the application of the NTA, contemporary "nontraditional" work may be part of the tradition and custom of a particular community if its use of knowledge is "traditional."

It is relevant here to consider the case of indigenous Australian artist Leah King-Smith, whose work appropriates the ethnographic images in early colonial photographs of the exotic Aboriginal other and represents those images using techniques of cibachrome and rephotography. In this way, King-Smith challenges the immediacy and objectivity presumed by this early anthropological arrest of the "ethnographic present" and the way in which the subject was confined to a particular geohistorical moment in colonization. This work was controversially rejected by the art committee of the Cologne Art Fair as lacking authenticity, criticized as contemporary art that merely followed the tradition (Papastergiadis 1998: 90).

Although the decision to exclude King-Smith's work from the exhibition was reversed after a very public outcry, it is a key example of the tension between the frozen "traditional" or geographical historical moment and the "unoriginality" and therefore unprotectable nature of contemporary interpretations of one's own culture. Thus, the work is beyond protection both from the point of view of the very limited historical and sentimentalized view of community, as well as the restrictive regime of intellectual property. Furthermore, this case demonstrates the invalidity of a system that values the object of protection according to originality and inventorship, in that such a system cannot recognize value in the process of transmission by tradition, of belonging, and of value derived from and constitutive of the community rather than the individual personality of intellect, or of innovation that travels demonstrably beyond that which has passed before.

In her transmission and continuation of community knowledge and her self-recognition and mutual recognition through that process, King-Smith rendered her work "unintellectual" as it were, in that it did not persist as an individualistic self-conscious innovation upon or derivation from what had gone before. Instead, it was effectively rendered mere imitation by the Cologne art committee, in their blindness to the mutual or circular relationship between individual definition and community integrity. Thus, the creations of King-Smith could not be reconciled as art according to the strictly linear progress that attends Western legal notions of originality and intellectual creation and the exclusive rights created by intellectual property laws. Indeed, requirements of originality and invention logically undermine the nonlinear differentiation and customary and cultural dissemination within indigenous community groups (Papastergiadis 1998: 90).

If intellectual property systems necessarily and strategically simplify creativity and the process of cultural knowledge in this way, how might the application of such laws to traditional knowledge protection be anything but unjust? In other words, through the application of intellectual property frameworks to traditional knowledge production, the "relations" of community are necessarily translated to the narrative of innovation required for intellectual property to be sensible. Thus, what would be protected might not necessarily capture the interests of community but rather the translation of traditional knowledge into intellectual property.

Community as the Circulation of Relations Rather Than Historical Entity

What is required is a community-based system of protection and custodianship, operating upon a "definition" of community according not to the constitution of the group but to the values and principles of the relations as asserted from within the community itself:

> Thus the development of the theory of community places central importance on the actual principles that embody communitarian relationships and must be careful to avoid the prescription of the specific communities in which these bonds exist. If this can be achieved, then there is no need to engage in the often exclusionary and ethnocentric practice of selecting which communities are particularly worthy of the name (Little 2002: 65).

Despite the serious attenuation of community relationships under the NTA, this interactive rather than historical concept of community, as the relationships between individuals rather than the historical constitution of the group, is not unfamiliar to the Australian legal system. Although now abolished, the Australian Aboriginal and Torres Straight Islander Commission (ATSIC) was the background to important judicial application of traditional and indigenous community, in the interpretation and application of the national legislation of 1989 governing ATSIC– the Aboriginal and Torres Strait Islander Commission (ATSIC) Act.[23]

In applying this Act, the Federal Court of Australia has stated that communal recognition and integrity of the community is of paramount importance in deciding the question of Aboriginality. Drummond J has noted that while biological descent may be relevant, it is not sufficient to establish Aboriginality, and stated that communal recognition as an Aboriginal person is the best evidence available to prove Aboriginal descent.[24] Thus, the ATSIC Act presents a dynamic and organic model of community that is built upon the subjectivity of the relations between individuals rather than on objective indices of community such as geographical location.

Unlike the serious limitations seen in the NTA, the courts have identified Aboriginality as extending beyond a geographical and historical connection when applying the ATSIC Act. This administration of self-identification and mutual recognition is important toward developing a legal subjectivity for community:

> Aboriginality as such is not capable of any single or satisfactory definition. Clearly the Aboriginality of persons who have retained their spiritual and cultural association with their land and past will differ fundamentally from the Aboriginality of those whose ancestors lost that association.[25]

Importantly, the concept of Aboriginality cannot be legislated as such; that is, it cannot be subject to the imposition of formal memberships or regulated by institutional structures that would necessarily undermine the very concept of Aboriginality itself. In this same way, community must be self-regulated and self-determined; that is, the nonhierarchical and nonlinear organic nature of community must be facilitated by the framework created to protect traditional knowledge. Thus, the particular community will be entitled to exercise rights in respect of that material, according to customary law.

Community, therefore, persists through the mutual recognition of members, that is, the self-regulation of community. It is, therefore, a kind of social or political structure in which political and legal authority may vest, rather than a mere cultural group, as it were, to which the term "community" may be extended but within which there is no real sense of social interaction and cohesion. Indeed, a sense of "place," as it were, for groups dispersed through the forces of colonization may be found to persist through traditional management and interaction with resources, rather than through conventional geophysical or national boundaries. The "place" of community is the interaction between traditional and indigenous groups and their resources. Community is the "site" of culture.

Conclusion: Toward a Model Protection

Thus, community must be reconsidered beyond the notion of a spatial, geographical, and social manifestation of individual groups. Community must not simply exist as a particular location, that is, a projection of community onto physical, proprietary space. If a particular local community is to have effective *sui generis* rights to protect, manage, and limit its traditional knowledge within an international juridical order, then it must be a source of identity to which other "identities" in a civilized society owe obligations, it in turn owing obligations to those others. In other words, if community is to be given legal effect then it must become the subject of rights rather than a mere projection of historical and geographical identity (Bauman 2001).

If it is the local interactions that give effect to the autonomy of the particular community to which all individuals must refer, then the juridical order in which communities are to have capacity and authority must be a global order potentially beyond borders and interstate market contingencies to command the bargaining process, and beyond the evasion of such order through multilateral or bilateral trade agreements. To achieve authority and capacity as a legal actor, the community must have access to economic and legal systems through the international recognition of *sui generis* rights, rather than be generalized and moralized beyond a dialogue with the state.[26] In this way diverse indigenous communities would be able to express themselves in relation to their cultural resources and products, and the "community" in question becomes the object of political, cultural, and public interest. Unless any proposed international legal solution is able to provide the opportunity to assert difference as a community, the unique claim of indigenous communities with respect to resources will not be recognized, and the obligations enriched by these inclusive systems of custodianship will be tragically inconceivable.

Notes

1. The term "traditional knowledge" is taken to include genetic resources, and traditional knowledge in agricultural, medicinal, and other knowledge and methods, as well as folklore in works of art, performance, stories, and so on.
2. The WIPO IGC was established in the 26th Session (12th Extraordinary Session) of the WIPO General Assembly, held in Geneva, from September 25 to October 3, 2000 to consider and advise on appropriate actions concerning the economic and cultural significance of tradition-based creations, and the issues of conservation, management, sustainable use, and sharing of the benefits from the use of genetic resources and traditional knowledge, as well as the enforcement of rights to traditional knowledge and folklore.
3. Solutions to the protection of traditional knowledge, in the context of consistency with intellectual property laws, are the subject of current discussions within the WIPO IGC. See the discussion of the work of the IGC in Gibson (2004, 2005). At the Seventh Session, staged in Geneva November 1-5, 2004, the emphasis on consistency with international intellectual property protection was challenged by delegates of many developing countries and indigenous groups, who argued that the emphasis should instead be on adapting intellectual property laws to make them consistent with appropriate and relevant protection for traditional knowledge. See the Report of the WIPO IGC Seventh Session, available at www.wipo.int/edocs/mdocs/tk/en/wipo-grtkf_ic_7/wipo-grtkf_ic_7_15.doc.
4. Traditional knowledge is frequently beyond conventional intellectual property protection due to its construction within intellectual property frameworks as being knowledge in the "public domain." This is extremely problematic in that the "public domain" is a legal construct, without comparable meaning in the customary laws of many indigenous and traditional groups. This issue of the public domain and traditional knowledge was of intense interest and debate at the Seventh Session of the IGC, attended by the author. See the report of the Seventh Session (note 3).
5. At the Seventh Session of the IGC, several delegations from developing countries and indigenous organizations raised significant concerns with the artificial categorization of areas of protection, ignoring the significant interrelationships between traditional cultural

expressions or folklore, traditional knowledge, and the tangible resources in genetic and biological material. See the report of the Seventh Session (note 3).

6. Reform of current systems of protection for indigenous intellectual resources in order to achieve protection anchored upon the concept of community is endorsed throughout reports by indigenous and traditional groups. Discussions of the special requirements of indigenous intellectual production have included the call for separate legislation to protect indigenous intellectual property, acknowledging the very different value of that intellectual interest for indigenous producers, as well as the inadequacy of conventional legislative protection of intellectual property. (See Janke 1997) which recommends a *sui generis* legislative framework that draws upon customary laws and communal systems of ownership and management to include and protect all forms of indigenous cultural and intellectual property, including ecological and agricultural knowledge (Chapter 18). Yet the concept of community remains a highly problematic feature of contemporary political discourse, compromised by the often insubstantial and vague application of the term in modern policy rhetoric (Little 2002: 24).

7. This is the way in which "collective" operates in conventional intellectual property law paradigms to satisfy the requirements of "joint authorship" or "joint inventorship."

8. Fleur Johns (1994: 178) cites Eric Michaels as arguing that the simple binary relationship of individual versus collective represents "some phony appeal to the primitive, or to a recently manufactured tradition."

9. Indigenous artist Bulun Bulun quoted in Golvan (1989: 348).

10. *Bulun Bulun v. R & T Textiles* (1997) 157 ALR 193.

11. This consideration of consistency, as discussed earlier, was of critical concern to the delegations at the 7th Session of the WIPO IGC. See also note 2.

12. Preliminary Systematic Analysis of National Experiences with the Legal Protection of Expressions of Folklore. WIPO/GRTKF/IC/4/3, 22. Available at www.wipo.int/meetings/en/doc_details.jsp?doc_id=153830. See also Janke (1997).

13. This is relevant when considering ATSIC, which goes some way toward providing the legal and social space for this opportunity.

14. (1992) 175 CLR 1. This awareness of the relevance and the subsequent application of customary law in evidence was continued in the cases of *Milpurrurru v. Indofurn Pty Ltd* (1995) AIPC ¶91-115 and *Bulun Bulun v. R & T Textiles Pty Ltd* (1998) 157 ALR 193.

15. The arbitrary categorization of knowledge in spite of the significant interrelationships in customary systems is a major concern for indigenous groups involved in the discussions of the WIPO IGC. See the report of the Seventh Session (note 3). See also the earlier discussion and note 4.

16. Citing *Rubibi Community v. State of Western Australia* (2001) 112 FCR 409 per Merkel J. Kirby J considered that it may be possible to protect cultural knowledge under the provisions of the act and queried the separation of native title rights to resources in petroleum and other minerals from other rights in the land (such as those arising through the historical use of ochre) based merely upon the view that they are minerals that are mined by modern methods.

17. Ultimately, however, Kirby J did not provide a finding in respect of minerals and petroleum due to the extinguishing effect of the Mining Act 1904 (WA) and Petroleum Act 1936 (WA). See Jagger (2002). The principle in *Ward* was subsequently applied in the Federal Court of Australia decision in *De Rose v. State of South Australia,* handed down in early November of 2002, where O'Loughlin J declined to protect the disclosure of spiritual beliefs and practices related to "places on the land" under the NTA, stating that such precedent has "made it clear that matters of spiritual beliefs and practices are not rights in relation to land and do not give the connection to the land that is required by s 223 of the NTA" (Para 51). With due respect to O'Loughlin J, *Ward* does not necessarily decide that spiritual beliefs and practices do not give connection to the land, but that the NTA does not extend to the use of those resources beyond controlling access to the land in which those cultural resources are situated or practiced. The majority decision maintains

that intellectual property laws or fiduciary duties may afford some protection to these rights. What the decision in *Ward* does state is that claims to protection of cultural knowledge are not rights protected by the act. It is not in the joint decision but in the single decision of Callinan J that the rights to cultural knowledge are not considered to be in respect of land: "cultural knowledge does not constitute a native title right or interest 'in relation to land or waters'" (p. 393).

18. Page 246. See the International Covenant on Civil and Political Rights; International Covenant on Economic, Social and Cultural Rights. Article 12 of the Draft United Nations Declaration on the Rights of Indigenous Peoples provides the following in respect of the right to cultural knowledge:

Indigenous peoples have the right to practise and revitalise their cultural traditions and customs. This includes the right to maintain, protect and develop the past, present and future manifestations of their cultures, such as archaeological and historical sites, artifacts, designs, ceremonies, technologies and visual and performing arts and literature, as well as the right to the restitution of cultural, intellectual, religious and spiritual property taken without their free and informed consent or in violation of their laws, traditions and customs.

Of particular interest in this respect were the statements of the delegation of the Tulelip Tribe at the Seventh Session of the WIPO IGC, which called upon international human rights law and "higher" laws and the need to recognize the prior legitimacy of the customary laws of tribes with respect to their natural and cultural resources. For transcripts of these statements, see the Report of the Seventh Session (note 3).

19. Referring to the decision in *Yumbulul v. Reserve Bank of Australia* (1991) 21 IPR 481. Furthermore, Kirby J rejects the assertion in *Bulun Bulun v. R & T Textiles Pty Ltd* (1998) 86 FCR 244 that recognizing native title rights that are analogous to intellectual property rights would contravene the "inseparable nature of ownership in land and ownership in artistic works" under Australian common law. His Honour states that such a principle cannot be maintained where it offends justice and human rights.

20. As considered throughout, this question of "consistency" has been and continues to be a major topic of debate at the Seventh Session of the WIPO IGC.

21. Preliminary Systematic Analysis of National Experiences with the Legal Protection of Expressions of Folklore. WIPO/GRTKF/IC/4/3, 8 (see note 12).

22. Ibid.

23. At the time of writing, the Australian Federal Government had introduced controversial plans to dismantle ATSIC. The final abolition of ATSIC took place in March 2005. For a background to the ATSIC changes and review, see the brief released by the Australian government, *Make or Break?* (2003). See also the Public Discussion Paper, released in June 2003, *Review of the Aboriginal and Torres Strait Islander Commission* (2003) and the subsequent report released in November 2003, *In the Hands of the Regions: A New ATSIC* (2003). See also media on the controversy including the SBS news reports "Canberra March Protests ATSIC Abolition" (2004) and "Senate Inquiry Sought Into ATSIC Abolition" (2004). See also work of civil society organizations in relation to this plan, including ANTaR, Australians for Native Title and Reconciliation, at http://www.antar.org.au/atsic.html ("ATSIC: The End of Self-Determination?") and http://www.antar.org.au/atsic_lathamltr.html; and Friends of the Earth (http://www.melbourne.foe.org.au/barmah/barmah_sdaction.htm). Nevertheless, these changes to ATSIC do not affect the relevance of the jurisprudence of this act to the present discussion of community.

24. *Gibbs v. Capewell* (1995) 128 ALR 577.

25. *Shaw v. Wolf* (1998) 83 FCR 113 per Merkel J.

26. Moral authoritarian communitarianism is often associated with the principle of the state as an enemy of community (see Hughes 1996). See also the more extensive consideration of the relationship between the state and community in Little (2002).

References

Bauman, Zygmunt. 2001. *The Individualized Society*. Cambridge: Polity Press.

Bergson, Henri. 1944. *Creative Evolution* (1911), Translated by Arthur Mitchell. New York: Random House.

Charlesworth, Hilary, and Christine Chinkin. 2000. *The Boundaries of International Law: A Feminist Analysis*. Manchester: Juris Manchester University Press.

Donnelly, Jack. 1989. *Universal Human Rights in Theory and Practice*. Ithaca, NY: Cornell University Press.

Dutfield, Graham. 2004. *Intellectual Property, Biogenetic Resources and Traditional Knowledge*. London: Earthscan.

Gibson, Johanna. 2004. "Intellectual Property Systems, Traditional Knowledge, and the Legal Authority of Community." *European Intellectual Property Review* 26 (7): 280-90.

_____. 2005. *Community Resources: Intellectual Property, International Trade, and Protection of Traditional Knowledge*. Aldershot: Ashgate.

Golvan, Colin. 1989. "Aboriginal Art and Copyright: The Case for Johnny Bulun Bulun." *European Intellectual Property Review* 11 (10): 346-55.

___.1992. "Aboriginal Art and the Protection of Indigenous Cultural Righsts." *Aboriginal Law Bulletin* 2 (56); 5-8.

Hannaford, John, Jackie Huggins, and Bob Collins. 2003. *In the Hands of the Regions: A Report of the Review of the Aboriginal and Torres Strait Islander Commission*. Canberra: Australian Government.

Hughes, Gordon. 1996. "Communitarianism and Law and Order." *Critical Social Policy* 16 (4): 17-41.

Jagger, K. 2002. "Ward: Mining and Petroleum." *Native Title News* 5 (10): 170.

Janke, Terri. 1997. *Our Culture, Our Future: Report on Australian Indigenous Cultural and Intellectual Property Rights*. Canberra: Australian Institute of Aboriginal and Torres Strait Islander Studies, Aboriginal and Torres Strait Islander Commission (ATSIC).

Johns, Fleur. 1994. "Portrait of the Artist as a White Man: The International Law of Human Rights and Aboriginal Culture." *Australian Yearbook of International Law* 16: 175-197.

Little, Adrian. 2002. *The Politics of Community: Theory and Practice*. Edinburgh: Edinburgh University Press.

Maddock, Kenneth. 1988. "Copyright and Traditional Designs: An Aboriginal Dilemma." *Aboriginal Law Bulletin* 2 (34): 8-9.

Mowaljarlai, David. 1995. Interview. "The Law Report," ABC Radio National, 31 October 1995. .

Papastergiadis, Nikos. 1998. *Dialogues in the Diasporas: Essays and Conversations on Cultural Identity*. London: Rivers Oram Press.

Perkins, Rachel, and John Romeril (screenwriters). 2001. *One Night the Moon*. Directed by Rachel Perkins.

Puri, Kamal. 1993. "Copyright Protection for Australian Aborigines in the Light of Mabo." In *Mabo: A Judicial Revolution*, ed. M. A. Stephenson and Suri Ratnapala. St Lucia: University of Queensland Press.

Underkuffler, Laura S. 2003. *The Idea of Property: Its Meaning and Power*. Oxford: Oxford University Press.

"Talking About *Kultura* and Signing Contracts"

The Bureaucratization of the Environment on Palawan Island (the Philippines)

Dario Novellino

In the Philippines, legislative requirements and bureaucratic procedures associated with indigenous claims over land and resources do not create a facilitating environment for sustainable culture and community-based practices. The old, strictly punitive protectionism is now being replaced by equally dangerous "community-based" forest management programs. Indigenous communities are no longer evicted from their territories; instead, they are asked to enter into agreements with the state. In spite of their promising features, community-based forest management agreements (CBFMAs) may contribute to the erosion of community livelihood and social cohesion while having adverse effects on fragile forest ecosystems.

Today, the Batak of Palawan[1] are inescapably trapped in a "state discourse" on property rights and environmental protection, with which they have great difficulties coping. Furthermore, they are all too aware of how state bureaucracy cannot be challenged through straightforward descriptions of their knowledge and beliefs. The latter, in turn, cannot be translated into the language of bureaucracy (Novellino 2003b).

This essay looks at these issues through an analysis of the Batak notion of *kultura* and the way in which this category is negotiated and discussed by the people themselves. Overall, *kultura* has been endowed with new systems of meanings and continues to be subject to multiple interpretations. Such meanings may include the adoption and transformation of state ideology by local communities, as well as the rhetorical use of indigenous values and notions by the state.

Batak reification of culture provides people with new and alternative means of articulating and verbalizing their experience and perceptions of "culture loss" and cultural revival. As Kirsch (2001: 168) has argued, "the concept of culture loss poses a problem of analysis for anthropologists given contemporary definitions of culture as a process that continuously undergoes change rather than something which can be damaged or lost." Yet, in my opinion, the use of the notion of "culture loss" is justified in a context where the people themselves talk about *kultura* using "mechanistic" metaphors that appear to objectify culture, and to compare it to something that can be "left behind" (*indi*), from which one can "separate" (*bilag*), that one could "go back to" (*balik*), that can be "removed" (*mairi*) and even "damaged" (*ranga*). Indigenous statements capture too well Batak attempts to cope with a multiplicity of forces and circumstances larger than and beyond themselves. In the second section, I investigate the complex set of relationships that entangle the Batak in the state and the consequences of peoples' attempts to receive state legitimacy over their land and resources and to carve a niche in a rapidly transforming world.

"Talking About *Kultura*"

The notion of *kultura* is spreading spontaneously among the Batak, as well as among other sectors of the Filipino population. This notion, as the Batak employ it, retains a high degree of flexibility and can assume different meanings according to the context in which it is used.[2] For instance, talks on *kultura* often provide a set of images through which people reflect and organize experiences of culture loss and their decreasing ability to make a living.

It is especially important to understand the meaning of *kultura* now that the Batak are beginning to negotiate their own discourses on culture loss with a wide range of outsiders. Interactions with outsiders require the utilization of terminology and concepts that are familiar to all interlocutors engaged in communication. It follows that local terms for custom, such as *ugalin* and *lai'*, may become of secondary importance, while new substitutive terms, such as *kultura*, are perceived as more powerful and effective when engaging with outsiders. Interestingly enough, the Batak speak of *kultura* to communicate not only with outsiders but also among themselves, for instance to describe their relation to the past, the transition that has led to the loss of their traditional practices, and the tension between their ideal projections of culture and their present lifestyle. Moreover, through the notion of *kultura*, the Batak express their views and sense of community that vary from one generation to the next. As a result, it is difficult to achieve agreement on how to counteract or cope with community fragmentation, loss of social capital (Novellino 1997), and loss of natural resources and cultural practices. Similarly, it is difficult to formulate adequate claims for the things that have been, or are being, lost. (See Figure 1.)

Figure 1. *A group of Batak on their way to Kalakuasan.*
The Batak have lost access to coastal areas because of the national road and
encroachment from migrants, March 2001.

Custom (Ugali)

The Tagalog (Philippine national language) term *kaugalian* can be translated as "shared customs, conventions, or deportment" (McDermott 2000: 219). According to Melanie McDermott, this was the term used by the Batak and Tagbanua of Kayasan in conversation with her, and often to refer to "traditional customs" (e.g. marriage, exchange of labor, food or land, reciprocity), conflict avoidance: and the avoidance of migrants. The Batak words for "custom," however, are *ugali* and *lai*; often these words are used interchangeably, although the latter is more commonly employed to stress ethnic membership and affiliation. Generally, to have a "good *ugali*" is the equivalent of displaying good behavior and a good attitude (e.g., being sociable, generous, and humble). To be in accord with each other is often perceived as the equivalent of "agreeing in customs" (*maguyunan it ugali*). Those who behave in a customarily accepted way (e.g., share food, respect elders) are said to possess good behavior/custom (*magayen a ugali*). Other customary practices (e.g., wearing bark cloth, anklets, armlets) are also regarded as an expression of the Batak *ugali*. However, it is important to point out that when Batak discuss issues related to their sense of community, "culture loss," and the revival of traditional practices, they seem to employ, more frequently, the word *kultura*. This is a term of Spanish origin that has been absorbed into Tagalog.

Kultura

The meanings conveyed in notions such as "custom" and *kultura* are based upon how these words are employed in various discursive forms. In the Batak language, the word for "removing" is *bawas,* which generally refers to reduction in quantity and value, as in removing water from a container or bringing a price down. However, with reference to the loss of certain traditional practices, I once heard a Batak using the following expression: *"kultura ta nabawasna gtiek,"* literally "our *kultura* has been removed a bit" (i.e., certain practices have been lost), and other people saying that, compared to the past, *kultura* was now "short" (*diput*) and shallow (*lubaw*). However, something could have been done "to add more" (*dugang*) to *kultura* (in the sense of bringing back practices that have been abandoned). On another occasion, I recorded a statement from a former Batak chieftain, Timbay, expressing his disappointment toward the other community members who, according to him, "were sleeping over their own culture" (*nakaedep at sadilin' kultura*). He added that he "awakened" (*masulagna*) from this "sleepiness" and decided to separate from the group so that Batak *kultura* would be "alive" (*egen*).

The lack of unity and declining in-group solidarity is one of the most common subjects associated with both loss of *kultura*, the decreasing ability to make a living and a failure to replenish natural resources. Often, the Batak themselves attribute the breaking of social norms to the acquisition of new habits from the outside. This is how Yolanda, a Batak woman in her forties, expresses the loss of solidarity. She claims:

> This is the reason why, little by little, our *kultura* has been removed [*mairi*]: too many new teachings have entered into us. For instance, what the pastor says is also acquired by us. The Kristianos [*Filipinos*] do not share food, and even when they have meat, they'll keep all of it for themselves, so now our people follow the same habit.

The Batak are becoming aware that increasing participation by their children in the educational training organized by missionaries or government agencies is vital to the learning of reading and writing. On the other hand, children's increasing participation in alien educational curricula is having negative repercussions on the transmission of traditional values and practices. This idea is clearly expressed by Timbay:

> This is why our *kultura,* little by little, has been removed. Look at our young people today. They are ashamed to wear bark cloth because the Filipino girls laugh at them. So they wear trousers; now they are civilized [*civilisado*]. That is why it is so hard to educate our children according to the old customs. If we insist, they will run away from the house. So we must find a balance. The more young people become "educated" [receive formal education] the more they separate from their own *kultura.*

There is an apparent tension between Batak attempts to secure a formal education for their children and the realization that this may represent a further agent of deculturation. While the Batak become increasingly exposed to external teachings, the young community members do not learn shamanism. Community members often express critical views on the abandonment of old practices. According to Pekto, a Batak from Tanabag: "it is the fault of the elders. They are the first to abandon old practices. So they do not pass them to children; now *kultura* is shallow *[lubau]*." Lito, a Batak in his late twenties, also blames his fellow villagers for the "loss" of culture: "the government cannot be blamed for the abandonment of our swidden practices. Although they forbid us, nothing will prevent us from planting rice, if we really want to."

Another common argument is that Batak culture will not be lost until people cease to speak the vernacular. For Pawat (a man in his mid-forties): *ampang magbaya it kultura* –literally, language accompanies *kultura*–and "as long as language is retained *[quitan]*, *kultura* will not be lost *[da'gwa nalipatna]*." On the contrary, "those who cease to speak their language have separated *[nagbelag]* from or left behind *[indi]* their *kultura*."

Apparently, the Batak fail to recognize that many imported words (e.g., *kultura*) are now acquired and mastered by elders and youngsters, shamans and nonshamans. Overall, Batak acquisition of new words such as "sustainability," "protection," and "development" has increased proportionally with their participation in the state. One could not communicate with politicians, or even understand the radio, without knowing the new words.[3] New words facilitate communication in the same way that a rifle (rather than a traditional bow and arrow) facilitates the catching of game animals. In addition, new words can be displayed communicatively with pride, in the same way that a radio is displayed as a status symbol. Batak who acquire new words are most keen to use them during skillful conversations with other village members. Speaking more languages and knowing more words is something to be proud of.

The arrival of tourists, especially during the *lambay* honey ritual ceremony, is another tangible sign of the ongoing transition (see Figure 2). This has encouraged the Batak to make sense of the new intrusion and to understand why their *kultura* is appealing to others.[4] On different occasions, I asked Batak what they think about tourists and why the tourists are visiting them. According to the shaman, Padaw:

> The tourists want to see what the "real Batak" looks like. For sure, they would not come all the way here if we would look just like the Kristianos [e.g., Filipinos]. They like to see us wearing bark cloth, but today we wear trousers. Yes, when they come here we are wearing trousers, but when we go to the city [to attend government-sponsored cultural events] we wear bark cloth.

Figure 2. *Women sharing the first harvested beehive during the lambay,*
March 1993

The aesthetic of Batakness is another local topic of discussion. Once Pawat told me: "the tourists need to see our pubic hairs to understand that we are Batak."[5] On August 21, 1999, during a village meeting, Ubad, the oldest person in the community, took a provocative stand on what an authentic Batak should look like and said:

> I think that if we really want to retain our *kultura,* if we want go back to the old way, then we must throw away our blankets, our trousers, and our tee shirts. We should sleep around the fire when it gets cold. My hairs are white, I'm an old man now, and if all of you agree with what I have said, I am more than happy to go back to the old ways. However, if all of you will continue to wear trousers and tee shirts, then you are a Filipino now … we can do nothing about it.

Ubad's statement was intentionally "extreme" to provoke the reactions of the participants. In reality, other Batak I have talked to do not perceive the wearing of Western clothes as a negation of their identity. For instance, says Katalino: "when I am with the Filipinos, I can be like a Filipino. When I am with the Batak, I behave like a Batak."

More recently, because of increasing contacts with tourists, the Batak have begun to think about *kultura* in monetary terms. The "commoditization" of *kultura* may soon encourage the Batak to draw a direct connection between loss and compensation. As of now, the Philippine legislation does not provide significant options for the compensation of indigenous land, resources, or other rights that have been extinguished through the implementation of estate companies, plantations, mining, logging, and so on. On the contrary, Presidential Decree No. 705

specifies: "when public interest so requires, steps shall be taken to expropriate, cancel defective titles, reject public land applications, or eject occupants thereof."

While state compensation for indigenous rights is still lagging behind schedule, I suspect that, in the near future, the revival of Batak traditions will be increasingly related to profit-making initiatives. As of now, the revivals of old practices are viewed with skepticism, especially by members of the younger generations. Even elders agree that it is difficult to reestablish the balance between communities and their environment, because certain resources are now perceived as mere commodities. This concept is clearly expressed by Padaw:

> Since the time of our ancestors all the products from the forest were for our consumption, not for sale. God gave all these things to us, so that we could be alive and never starve. Then we began to sell the meat of the wild pig, the wild honey, and other things of the forest, so the Master of Pigs and Master of Bees became upset. It was then that our *kultura* began to be destroyed [*ranga*].

In the Batak language, to forget and to lose are synonymous, and thus loss does not necessarily lie in the abandonment of a practice. In fact certain practices, such as the making of bows and arrows, could be replicated since a few people in the community still have the knowledge of this technology. However, these practices are no longer relevant to the present circumstances and, in this sense, are said to be forgotten.[6] The hope that certain customary practices will "return" is not completely lost, as long as such practices are remembered and made relevant to present needs. In the Batak language "to return" (*balik*) often denotes the physical action of going back (e.g., to one's own village), but it can be employed to indicate the return of practices that have been lost or are being abandoned.

On August 10, 1999, a group of Batak gathered together to discuss these topics. On that occasion Ernesto, a Batak from Tagnipa, claimed: "if we talk about Batak *kultura*, I really want my children to learn it. But if we talk about going back to our old *kultura*, then I am unsure about the things we should regain. Can any of you share his opinion on this?" The reply came from Elisio. He said:

> In my opinion it is like this. This is the *kultura* that I would like to see coming back: meat of wild pigs shared among people, not sold to the restaurants. All of us should continue to make *tarakabut* [inviting people to harvest the first mature rice ears]. If you see that your neighbors have nothing to cook, share your food with them. This is what the ancestors told us. These are the things that should come back.

The reply from Elisio placed emphasis on "good behavior." Instead Padau, in response to Ernesto, stressed the value of actual "livelihood" practices. He claimed:

> In the same way our forefathers did things in the old time, so we should do them now: *lugitem* [searching for food in the forest], *lambay* [performing the annual honey

ritual], *da'es* [sleeping in temporary camps along rivers during the dry season], *maglugu* [poisoning fish with toxic plants], *uma* [swidden cultivation]. These are the things they did. We still do some of them, but on a smaller scale …little is left.

Remarkably, the people tend not to see their involvement in traditional livelihood and ritual practices in contradiction with attending the Sunday mass or other church-related activities. This is echoed in the words of Elisio:

When it comes to custom [*ugali*], I am a Batak, my mother is a Batak, my father is a Batak. It is true that I am now a Christian [converted to Christianity], but this is only in relation to my religion; it is something that concerns God. But when it comes to my *kultura*, I take part in it, because I am a Batak. This is what the pastor told me: I will teach you the words of God, I have no intention of destroying your *kultura*, because your *kultura* is yours, it belongs to you.

Batak "religious" practices are intertwined in every aspect of daily life, and the Batak have no word to define religion as a specific domain separate from others. The Batak realize that "lack of religion" may limit their participation in the socioeconomic development of the state, while membership in a religious sect may allow them to counter discrimination.[7] Today, Batak and migrant Filipinos join together in various religious activities. However, the former are also aware that to join a religious sect may constrain their participation in shamanism. So, at least in principle, a perceived alternative to safeguard both traditional beliefs and their participation in a wider society is to separate religion from culture. As Elisio suggests, to belong to a certain religion does not necessarily imply the separation from one's own culture. In the same way, to join a shamanic séance is not seen in contradiction to one's own affiliation with a Christian sect.

Apart from joining religious sects, the Batak may avail themselves of other concepts borrowed from the national language, such as *katutubo* (indigenous people) and *tribu* (tribe), to perceive themselves as being part of a larger network sharing a similar economy and "lifestyle." The Batak recognize that the term *katutubo* includes people whose *kultura* can differ in various ways, and yet this notion and other terms, such as *tribu*, can be used to stress similarities and a common livelihood with other groups.

State Rhetoric of Kultura

The way in which the Batak rely upon the notion of *kultura* to express "culture loss" and identity could not be fully understood without taking into account state rhetorical use of *kultura* and the impact that this has had on Batak ways of thinking. I have selected two statements extracted from speeches given by a local missionary and a political candidate. Even a cursory glance at such statements indicates that Batak and the dominant society's views of *kultura* present overlapping features.

A politician running for municipal mayor made the first statement. He visited the area in March 2001, while the community was busy with an annual ceremony. One extract from his speech goes as follows:

> I love the *katutubo,* we are like brothers, we all share the same land, and we all want a bright future for Palawan. I respect your *kultura,* your traditions. You can wear your G-string and visit my office, and this will be okay. You must be proud of your *kultura.* If I win the election, I will allow you to do *kaingin* [shifting cultivation].

The following statement was recorded in April 2001, in the local Church of Kalakuasan, and was part of a speech given by a Filipino missionary, now living with the Batak community. He said:

> You can become a Christian and still retain your own *kultura.* God gave to each group its own *kultura,* and he loves all his peoples in the same way. You can bring your drums, your musical instruments to the Church … You are free to express your *kultura* inside the church if you want to. Yes, come to the Sunday service, wear your own clothes, bring your tools, do not be ashamed of your *kultura.* In the eyes of God we are all the same.

Clearly, both rhetorical styles portray the church and the state as benevolent entities that recognize and even celebrate cultural differences.[8] The underlying state rhetoric is that all ethnic groups with their distinctive *kultura* are an expression of an all-encompassing Filipino identity; they are all part of a single nation with a unified economy, and identical rights and duties. In the eyes of God and/or the state everyone is valued in the same way, independently of his/her culture. The Batak have now become aware of the local rhetoric and do realize that the state is acting according to its own agenda, rather than in the public interest of those having a *kultura.*

Signing Contracts

In 1997, the Batak of Tanabag returned to the permanent settlement of Kalakuasan. They were encouraged by the local government to do so and were told that, by moving closer to the seashore, they would become citizens with the same rights and responsibilities as the people living in the coastal villages. After resettling in Kalakuasan, Batak responsibilities toward the government (attending meetings, registering for elections, etc.) increased, but the government has not made apparent improvements to enhance people's rights to land and resources. Conversely, the Batak environment has become increasingly bureaucratized, and state laws have made land and resources even more inaccessible to local populations. Living in a permanent settlement has made the Batak totally vulnerable and "locatable." By "being locatable, local peoples are those who can be observed, reached and manipulated as and when required" (Asad 1993: 9). Unexpected visits by politicians are also time-consuming. Especially during election time, candidates may visit the Batak without previous notice, and yet they expect people to attend their speeches. Alternatively, village members are requested to go down to the coast to welcome a certain politician, who may never show up. So the Batak will waste a full day, without receiving either food or drinks from those extending the invitation.

The Batak themselves perceive their involvement with the state cash economy as a main cause of poverty and decreasing control over land and resources. In addition, researchers from different backgrounds are visiting the Batak territory, and the people are becoming increasingly concerned about the impact of these intrusions. Specifically, the Batak want to see a practical value to what researchers do in their community, and how academic publications can be used to support indigenous rights to land and resources. Thus far, anthropologists' attempts to draw upon existing laws to protect indigenous people's rights are particularly problematic (see Weiner 1999).

The "Community-Based Forest Management Agreement" (CBFMA)

In spite of their lingering popularity, CBFMAs launched under the grandiose banner of "participatory approaches" have neither contributed to safeguarding indigenous rights over land and resources nor led to a more sustainable management of forest resources (Novellino 1999, 2000a, 2000b). CBFMAs belong to a policy of the Department of Environment and Natural Resources (DENR) that allows local communities to manage forests that have been converted to nontimber uses. One of its objectives is to develop self-sustaining production systems in the uplands by replacing indigenous swidden practices with permanent forms of agriculture. A closer look at CBFMAs reveals that such a policy violates indigenous people's rights to their ancestral land and perpetuates government control over indigenous people's lives. In fact, with CBFMAs, indigenous people's role in their own territory is reduced to that of stewards of the public land. For instance, in the agreement entered into between PENRO (Provincial Environment and Natural Resources Office) and the Association of Batak of Tina on December 18, 1998, it is specified that the indigenous beneficiaries should "immediately assume responsibility for the protection of the entire forest-land within the CBFM area against illegal logging and other unauthorized extraction of forest products, slash-and-burn agriculture [*kaingin*], forest and grassland fires, and other forms of forest destruction, and assist DENR in the prosecution of violators of forestry and environmental laws."

Clearly, the contract requires that the Batak themselves must guard their area from their own practices, such as swidden cultivation. CBFMAs do nothing to recognize the claims of indigenous communities over their ancestral domain. Rather, they place indigenous forest management under government control and use the people as subcontractors of the DENR. It is worth noting that members of the Batak community claim to be unaware of the real content of the CBFMA, which is written in English rather than in the national language, Tagalog.

The Livelihood Alternative Dilemma

On August 21, 1999, a forester from the DENR visited the community of Kalakuasan to obtain local consent for a number of livelihood and forest protection activities to be carried out in connection with the CBFMA. Again, the Batak were faced with the prospect of giving up their traditional swidden cultivation practices and being forced to accept livelihood alternatives such as aquaculture and "contour-farming," which are objectionable to them (Novellino 2003a). I had previously told my Batak friends that by entering into a CBFMA they were obliged to accept all conditions of the contract, including the prohibition on swidden cultivation. Therefore, by signing that contract, they would have limited the success of future actions to support their right to practice "traditional" upland farming. Apparently, the Batak were not particularly worried about these matters. Rather, as they explained to me, their immediate concern was to obtain the legal authorization to gather and sell nontimber forest products (NTFPs) in their area. They also reminded me that, in spite of the agreement, they would have continued to open and cultivate swidden fields in certain locations not easily accessible to outsiders. (See Figure 3.) In other words, the Batak were willing to face the

*Figure 3. Batak resting after the planting of upland rice in the
remote interior of Tanabag, April 2004*

potential risk of being apprehended by forest guards, as long as they would have received legal rights to collect and sell NTFPs.

With the CBFMA in place, things have turned out to be much worse than the Batak had expected. As of now, the Batak have been unable to fulfill most of the bureaucratic obligations in relation to their CBFMA. Specifically, they did not submit their Annual Work Plan (AWP) and the Community Resource Management Framework (CRMF) to the Community Environment and Natural Resources Office (CENRO). It should be pointed out that these reports are to be written according to strict government standards, and the Batak, being illiterate, do not have the technical skill to prepare them. Because the Batak are unable to produce such reports, DENR has withdrawn the permits needed by the community to sell their NTFPs. In 2001, because of the state prohibition against swidden practices, most Batak families had already lost most of their traditional varieties of rice.

Between May and July 2001, I assisted the Batak in the preparation of the AWP and CRMF, in collaboration with a local organization. Both documents were prepared and formally submitted to the concerned authorities in May 2001. Preparation of these documents required close coordination between the community members and myself in order to discuss controversial topics, such as the inclusion of swidden practices in their AWP. I was pleased to see that the Batak were willing to challenge the government prohibition on shifting cultivation. On the other hand, they were concerned that the inclusion of swidden farming in the AWP might have led the DENR to take legal actions against them, thus jeopardizing the community's future chances for obtaining the necessary permits to gather and sell NTFPs. A unanimous decision was eventually reached: the Batak agreed that swidden cultivation had to be included as one of the activities of their AWP and that this decision had to be forwarded to the concerned government agencies.

To challenge DENR regulations was, in my opinion, a remarkable move. However, I thought that, in order to strengthen communities' claims to swidden cultivation, it was necessary to support and validate them, using powerful pieces of legislation. Thus the appropriate laws had to be identified and carefully assessed. Interestingly enough, Presidential Decree No. 705 prohibits shifting cultivation nationwide, while Republic Act 8371 ensures protection for indigenous rights to perform traditional religion. Significantly, swidden cultivation (*uma*) is not only a food-producing activity but also a fundamental part of people's ritual practices and "religious beliefs."

According to Sec. 33, Chapter VI of Republic Act No. 8371, "Indigenous Cultural Communities (ICCs)–Indigenous Peoples (IPs) have the right to practice and revitalize their own cultural traditions and customs." Moreover, as specified in Rule VI, Section 3 of the "Rules and Regulations Implementing Republic Act No. 8371," the right to cultural integrity shall include "recognition of cultural diversity"; "protection of religious, and cultural sites and ceremonies" (no doubt, this also applies to Batak swidden fields and related rice ceremonies); and "right to protection of indigenous knowledge systems and practices" (again, such rights should include Batak knowledge of local crop varieties and agricultural practices). In addition, Section 12 stresses that ICCs/IPs have the right to "manifest, practice, develop, and teach their spiritual beliefs, traditions, customs, and ceremonies." Clearly, all such rights were being hampered through the implementation of CBFMA regulations. The more I studied the law, the more I became convinced that this could have been used to support Batak claims to swidden farming. After a careful assessment of the existing legislation, I held discussions with the Batak about the fundamental connection between the sustainability of traditional swiddening, local beliefs, and ritual/religious practices, and on how the existing legislation might have been used to validate such a connection.

A Batak legend attributes the origin of rice to a human sacrifice. Each year, before planting rice, the people practice a number of ceremonies to call back the *kiaruwa'* (lifeforce) of the child who was killed by his father in legendary times. Germination and health of rice seeds is said to depend on the action of the "child's lifeforce." Each year, the Batak welcome the return of the *kiaruwa'* of rice, also referred as *kiaruwa'* of the child. For this purpose they perform a number of ritual activities in the center of the swidden field, and rice is generally regarded as *taw* (person/human).

Particularly promising, I thought, was to use Batak beliefs and rice-related practices as evidence to support people's rights to "protect indigenous knowledge systems" (and thus swidden cultivation). In other words, during our meeting with the government, we would have demonstrated that, according to Republic Act No. 8371, the state was obliged to protect, rather than prohibit, indigenous agricultural practices because they are also an integral part of Batak "spiritual beliefs." My argument sounded convincing to most Batak; however, some were unsure about the implications of disclosing the most profound aspects of Batak culture to government officials.

Finally, the meeting with the CENRO official was arranged. Two Batak representatives, members of a local nongovernmental organization (NGO), and I visited the CENRO office in Puerto to discuss and defend the argument that swidden cultivation had to be allowed inside the CBFMA area, and be regarded as one of the activities of the AWP, and of the CRMF. Contrary to what we agreed, during the meeting, the Batak did not discuss with government officials the connection between rice cultivation and religious beliefs, nor did they mention the myth concerning the origin of rice and why this crop is often referred to as *taw* (human). It was only in the following days that I became more and more aware of why the Batak are unwilling to disclose certain aspects of their culture to outsiders. In short, I had failed to see that the

Batak are all too aware that state bureaucracy cannot be challenged through "direct" and straightforward descriptions of people's worldviews. In relation to this, Pekto, one of the Batak joining the meeting, told me:

> How can we explain to the government that rice is human? I am sure that they cannot understand this; they would laugh at us. Exposing these issues would make things even more complicated. Because the government would ask us: do you have proof of what you say? Do you have a document to support what you say? The government is different from us. They always have a piece of paper for everything they say, but our *kultura* is only "on the tongue," we have no written papers, so we cannot challenge the government.

So far, my attempts to assist the Batak in fulfilling all the requirements listed in their CBFMA have not been met with success. CENRO decided not to approve the Batak AWP and CRMF because, according to them, some of the required information was missing. Such information includes technical details that not only the Batak but even I have difficulties comprehending.

Also, our attempts to prepare a participatory map of the Batak territory did not meet "DENR standards." Irrefutable proof of original occupancy, I was told by the Batak, lay in the legend of "Esa'." He was the ancestor who gave a name to all places in the Batak land. Thus, we included in the map the local names of streams, mountain ridges, and mountain divides that, according to the Batak, were given by Esa' in mythological times. In the same map we gave some indication of the areas cyclically used by the people for their swidden cultivation practices and the area where the resin of *Agathis philippinensis* is gathered.[9] (See Figure 4.) CENRO argued that the map was technically incomplete, and they insisted that we also specify the places where rattan resources were located and

Figure 4. A Batak tapping an Agathis philippinensis tree, May 1999

all the fields that the Batak were planning to cultivate in the next five years. Again we explained that this was impossible. First, we argued that we could not highlight any specific location for rattan climbing palms, since these are found everywhere in the Batak territory. Second, we explained that it was impossible to predict the future location of individual swiddens, since the decision to recultivate certain areas depends on marriages, deaths, and intracommunity alliances that we could not anticipate. CENRO officials were not satisfied with our explanations, and again they asked us to comply with the law.

After a few days, CENRO sent a group of foresters to Kalakuasan to settle a boundary dispute between the Batak and their Filipino neighbors. During a meeting, the foresters asked the Batak to point out the boundaries of their area on a technical map simply made of straight lines and topographic symbols. The map bore no observable natural features of the landscape and was thus incomprehensible to the Batak. The request of the foresters caused much indignation amongst the Batak participants. Pekto, the chairman of the CBFMA, stood up and said:

> We cannot understand what you say; we cannot understand your map. You only draw lines; you don't mention the names of the mountain ridges, the names of rivers, the names of trees. If you mention the mountain divides, the streams flowing into the main river, we can understand you. These are the things we know, so even when we walk in the night we know where we are going. Your map, your definitions tell us nothing!

Foresters' efforts to invalidate Batak claims through the use of their own technical maps involved a deliberate attempt to create fixed boundaries and contrast between inside and outside that is alien to Batak perceptions of landscape, but–surprisingly–it features in their description of and discussions about *kultura*.

Discussion and Conclusions

One of the arguments arising from this ethnography is that the pursuit of *kultura* is "part of the search for identity in the contemporary world" (Sillitoe 2000: 241). This notion is becoming a powerful tool for the Batak's engagement in discussions about themselves vis-à-vis non-Batak others, and as a means for articulating the dialectic of present and past and the dynamics of loss. The important point is that when the Batak are reflecting on good manners and proper behavior, they often use the term *ugali*. When they project an idealized version of themselves, or discuss culture loss in comparison to past and present, they use *kultura*. It would appear that the notion of *kultura* not only overlaps with that of *ugali* but also incorporates it. The former seems to belong to a more comprehensive semantic field where meanings are negotiated and always open to contestation–allowing, for example, religion to be perceived as separate from cultural practices (Elisio).

At other times *kultura* is viewed as something that "cannot be removed" as long as it is remembered; as some sort of aesthetic expression, such as wearing or not wearing certain things (Ubad); something that is under threat, from which one could be separated, as well as something to which one could go back (Timbay), and that can be retained, as long as the local language is not lost (Pawat). At the same time, the ability to retain one's own *kultura* is also perceived as depending on people's capacity to keep a good relationship with non-human agents, for example, Master of Wild Pigs, Master of Bees, (Padau), and this is perceived as the precondition for a correct and "sustainable" use of natural resources. Thus, as the shaman points out, the commercialization of wild pig meat and honey has affected the relationship between human and nonhuman entities, and this has had negative repercussions, not only on *kultura* but also on the availability of vital resources. Customarily, the role of the shaman as mediator between the tangible and intangible worlds is essential to the maintenance of the social/cosmic order, and this has significant implications for the management of the environment. The continuity of cooperative behavior and reciprocal exchange, and the maintenance of good relationships with the other entities are all associated by the Batak with what is perceived as a "sustainable" use of the common environment shared with animals, plants, and nonhuman agents (Novellino 2003b).

Regaining one's *kultura* is also connected with the capacity to rebuild in-group solidarity and customary norms of behavior related to food sharing and labor exchange (Elisio). Conversely, the "removal" of such norms is viewed as a cause of culture loss (Yolanda) and risks putting "sustainability" or ecological balance in peril. The notion of *kultura* also includes food-seeking practices, related ceremonies, and patterns of mobility, as they have been inherited by the ancestors (Padau). Moreover, Batak *kultura* is perceived by the people themselves as "on the tongue" (oral). And yet, some Batak see the value of "putting their *kultura* inside a book."

Being "context-dependent," Batak discourse on *kultura* and loss is always enacted in an innovative and improvised way.[10] On certain occasions, *kultura* seems to be discussed by people as a "resource" or a "box of practices" where things can be taken out and placed in, or as something that can be either "increased" or "reduced," possessed or lost. Foresters regard uncultivated land as vacant and as something that can be mapped, measured by counting, and identified by the resources found within it. These approaches condense a particular Western notion of landscape that privileges a physical distance (Olwig 1993) and that views space as something that, in order to exist and be valued, must be measured and filled with things. A similar idea is reflected in the implementation of conservation-development projects, which tend to view the target communities as recipients of external assistance, and hence the whole village as an ideal space to be filled with environmentally sustainable initiatives. As Appell (1988: 274) has argued, "a too common characteristic of development projects is a tendency to view the target society as an empty vessel to be filled, not realizing that every introduced activity displaces an extant, indigenous one, possibly more critical to the survival of the target population."

I would argue that in modern Western thought not only space but also time becomes valuable when it can be filled with things. In fact, government officials often judged the Batak as lazy because "they make no good use of their time" and, as a DENR official told me, "they spend too much time roaming around in the forest, rather than investing it in reforestation and stable agriculture." Those of us brought up in the Western tradition would have no particular problem with the idea that "time must be used" and "space is there to be filled." However, I find it remarkable that these strands of thought have not gone unnoticed to the Batak. One would suspect that the people have made sense of these notions to the extent of incorporating them in their own representation of *kultura*. When *kultura* is imagined as a container (a box of practices), it acquires an inner and outer layer, a "depth" and a perimeter within which selected traits can be placed. In this way, *kultura* can be filled with anything that the Batak find useful to affirm their own identity in a specific context and at a particular time.

Clearly, the Batak use of *kultura* seems to suggest the existence of a fixed supply of practices and cultural features that can be turned to for help, support, or other purposes. Overall, when compared to *ugali* (custom), *kultura* offers new ways of expressing and articulating discourses on loss and may have good potential for facilitating the formulation of claims to land and resources. It is my impression that the notion of *kultura* presents the Batak with an image of themselves that renews the hopes that those things that are "taken away" (*mairi*) from *kultura* can be put back at some stage.

By stereotyping and polarizing the contrast between old and new (e.g., wearing or not wearing G-strings, sharing or not sharing meat), the Batak set past and contemporary experiences not in a mutually exclusive fashion but in a way that suggests the possibility of rehabilitating "old traditions," should the people decide to do so. Again, agency is displaced on people; they portray themselves as coresponsible for "removing" or "helping" their *kultura*, as well as for renewing or depleting natural resources.

The ethnography further suggests that the Batak do not always and explicitly verbalize the association between culture and knowledge, as one would generally assume. In the Batak language, *karunungan* is the most inclusive term to indicate skill, ability, talent, and practical knowledge. Different definitions are used to express the notion of "knowing how," and this includes a number of features of competence that are not made explicit when people discuss *kultura*. Instead, through the notion of *kultura* they often highlight stylized behavior, aesthetics, and form (i.e., behaving and looking like a Batak) rather than the skills that one should master in order to be a Batak. It is perhaps not surprising that, today, Batak youths are ever less concerned with demonstrating their Batakness in terms of traditional skills. The younger generation do not seem to perceive the abandonment of traditional hunting techniques, agricultural practices, and so on as especially dramatic or upsetting.

The line of argument taken so far suggests that individual variations on the *kultura* theme should be understood as one of the people's attempts to make sense of inevitable contradictions in their lives while being bombarded by all sorts of alien (and alienating) forces. As we have seen, the domain of meanings of *kultura* (the official state version) includes the set of inherent cultural rights that are allegedly recognized, respected by, and incorporated within the state. Overall, the Batak, missionaries, government officials, and NGOs all talk about *kultura* in ways that are similar and at other times different, and yet a discourse is formed that is open to multiple manipulations and misunderstandings.

Missionaries and politicians use the word *kultura* for their specific needs, but the Batak have their own agenda too: (1) to make sense of loss, especially when this relates to the exhaustion of natural resources and the deterioration of people's solidarity and trustworthiness; (2) to give order to a complex set of cultural features (which could not be simply conveyed through the term *ugali*); (3) to negotiate ways of expressing culture loss on more intelligible grounds with both governmental and nongovernmental organizations; and (4) to negotiate *kultura* in a strategic way to meet outsiders' expectations.

When Batak talk about *kultura*, they seem to be less interested than I was in the mystical of origin of rice and in the return of the "rice lifeforce" to their fields. This is because shamanism is increasingly seen as an irrelevant instrument to face and deal with the new transformations. Especially for the young generations, imported music and modern dances have an aesthetic force and a socializing dimension that largely overrides that of shamanic séances and traditional narratives. In short, the past offers them no clues on how to survive in a rapidly transforming world. Rather, it is perceived as a hindrance to participating in the wider Filipino society. Thus, the Batak are aware of the difficulties of using traditional beliefs as a means to legally support rice cultivation.

Given the drastic and continuous changes that the Batak are experiencing, the emergence of a culturally agreed upon and collective representation of culture loss is difficult to conceive, even in a small community such as Kalakuasan. Again, this brings us to the problem of how certain cultural values, no longer shared by communities as a whole, may be used by indigenous advocates as means to support customary rights (e.g., to rice cultivation) and to infer that local beliefs (e.g., the attribution of "personhood" to rice) play an essential role in the maintenance of both genetic diversity of rice and sustainability of traditional swiddens. Yet one wonders: if the knowledge that underwrites a belief system is no longer a basis for ritual action, can this still serve a political function for people like the Batak? (Weiner 1999: 203).

It is in this context that the issues of social cohesion, cultural erosion, and environmental sustainability acquire particular relevance. Today because of a lack of unity and various forms of exploitation by outsiders, the Batak are not developing counterstrategies as a group; rather they are forming a myriad of individual or family-based responses that are not always sustainable from an ecological point of view. For instance, the increasing reduction of game animals requires the acquisition of more sophisticated hunting devices. Moreover, dramatic environmental changes and adverse meteorological phenomena (e.g., El Niño and La Niña) have contributed to constrain Batak land use options and narrow their repertoire of sustainable resource strategies. The new trend is toward increased collection of valuable NTFPs that, however, is constrained by the number of non-indigenous gatherers encroaching within the Batak CBFMA area. Growing competition over NTFPs is forcing the Batak to contravene their own sustainable patterns of resin extraction. Hence, they are placed in a position that not only makes it impossible for them to reproduce their local knowledge but, paradoxically, forces them to infringe upon it.

The same applies to traditional swidden practices. Because of government prohibition against the conversion of forest into swiddens, the Batak cannot apply the customarily long fallow periods to their land. Rather, they are forced to cultivate soils that have not regained their nutrients and are still covered by bushes and small diameter trees. Notoriously, compared to plots cleared from long-fallow vegetation, these fields are particularly vulnerable to weeds. The people have to devote more time to controlling sun-loving grasses while experiencing dramatic yield decline. The government ban on swidden farming is also affecting the genetic diversity of cultivated plants. Some local varieties may become rare or even extinct if people are no longer allowed to cultivate them. Finally, the ban has placed an insupportable burden on the surrounding forest. This is because the local indigenous communities are now forced to increase pressure on NTFPs (e.g., *Agathis* resin, rattan, and honey) to compensate for the decline of agricultural produce.

As the sense of community slowly evaporates, the sustainable management of common resources is challenged by a number of factors that are complex and overlapping. In 2004, during one of my last visits, the community wanted to stop unauthorized gatherers from entering their CBFMA area but were afraid of intimidation and possible violence. Their legal authorization to gather NTFPs had expired and they had been unable to obtain a new one. Honey production had been unstable, and people had little cash to purchase rice. More importantly, several community members had contracted huge debts with buyers and, when the latter began to pressure them, the Batak could not pay them back–even the smallest amount.

Faced with this uncertain future, the Batak search for new places in the world. It does not matter how exploitative the relations with the state may be; the Batak will engage with it because the people have no other alternatives. Ironically, "being locatable," and thus controllable, is nevertheless a way of being in the world, rather than being excluded from it. On the other hand, as the population becomes more "locatable," and thus more sedentary, the feeding pressure concentrates on fewer areas with adverse repercussions on environmental sustainability. In turn, the transformation of both natural habitats and Batak social fabric is coupled with the redefinition of people's own identity and feeling of belonging. Surely, the notion of *kultura* is an instance of this, and it is consistent with people's attempts to craft a new sense of "community" and to frame loss vis-à-vis memories and contemporary practices of livelihood (Spyer 2000).

Remarkably, some of the local interpretations of the concept of *kultura* seem to contain an implicit claim to knowing and belonging to a large world: the nation. In turn, the Philippines operates as a nation by virtue of its rhetorical discourse on both cultural diversities and interethnic coexistence, as well as environmental sustainability. The latter notion is often used to merge the concept of "genetic erosion" with that of "cultural extinction." Because biodiversity is regarded as a public good, either indigenous people are expected to become "stewards" (not legitimate owners) of such good, or they themselves are transformed into a public commodity, a readily available resource for developers and project proponents. On closer inspection, it would appear that government attempts to promote "community forest management" systems have to do more with defending a particular notion of "the state as the supreme owner of the public land" rather than with a genuine interest in recognizing collective rights and interests. After all, the state ignores the complex aspects of its individual cultures and thus relies on inclusive notions such as *kultura, katutubo,* and *tribu* to forge a unity out of the country's remarkable human, linguistic, and religious diversity. On the other hand, Batak attempts to recast their own identity through the use of imported notions is just another manifestation of what people do, or are willing to do, in order to acquire state legitimacy (Spyer 2000). Whether such attempts will become an effective means to ensure environmental and cultural sustainability and to counter discrimination and injustices is, however, another matter.

Acknowledgments

This research was carried out while I was a Visiting Research Associate of the Institute of Philippine Culture of the Ateneo de Manila University. I would like to thank the Batak of Tanabag for their valuable support and friendship. I am grateful to Professors Roy Ellen and Bill Watson from the University of Kent at Canterbury for comments on an earlier draft. I acknowledge invaluable funding (grant no. 7136) from the Wenner-Gren Foundation for Anthropological Research.

Notes

1. The Batak are found scattered in the central portion of Palawan. They have a heterogeneous mode of food procurement, mainly centered on swidden cultivation integrated with hunting, gathering, and commercial collection of non-timber forest products. Eder (1987) estimated their population to be about 600-700 individuals in 1900, while his complete census in 1972 located only 272 with two Batak parents and 374 with one Batak parent (1987: 110). My provisional census in 2005 indicates that there are only 155 individuals with two Batak parents, a decline in the Batak "core" population of almost 57 percent within a period of thirty-three years. The present study concerns the Batak community living in the territorial jurisdiction of Barangay Tanabag. It consists of 28 families and an overall population of 126 members, of which only 68 individuals have two Batak parents.

2. Similar concepts such as *adat* in Indonesia and Malaysia (see Benda-Beckmann 1984; Ellen 1983, 1997; Geertz 1983; Hooker 1974; Josselin de Jong 1960; Murray Li 2000; Peletz 1988; Stivens 1996; Warren 1993; Zerner 1995) and *kastom* in the Pacific (Demian Forthcoming; Foster 1992; Jolly 1994; Sillitoe 2000; Strathern 1995) are employed by local people, state agencies, environmentalists, and so on to convey a wide array of cultural meanings, from religion, to land customary practices, to language and political institutions.

3. Tsing (1999:162) has argued that in Indonesia, the collaboration between village people, advocates, and policy-makers "give[s] life to concepts such as village development, tribal rights, sustainability, community-based conservation, or local culture ....these same concepts make political agency possible on both sides: they are the medium in which village leaders and those who study, supervise, and change them can imagine each other as strategic actors and thus can mould their own actions strategically." On the other hand, the use of these concepts entails that the people themselves are renegotiating their relationship with the forest and with "nature" more generally, and these conceptualizations depend "on how the people use it, how they transform it, and how, in so doing, they invest knowledge in different parts of it" (ibid: 139).

4. As Sillitoe (2000: 252) suggests, there is a profound irony underlying the relationship between tourists and local communities. With reference to the Sepik people, he argues that perceptions of what the outsiders want will determine the way in which people present their identities to outsiders. As a result the people will open for the tourists a window toward their past, a glimpse of their "primitiveness" before it changes inexorably in confrontation with the external changes of which the tourists themselves are a part. "The irony for the Sepik people is that what makes them to do this is the desire to earn cash that will allow them access to the products of the market–clothes, outboards, Coca-Cola–but success in this will undermine their capacity to continue earning because their attraction is as 'unchanged' and 'primitive.'"

5. Of the relation between tourists and local communities in Melanesia, Sillitoe (2000: 250) writes: "local people are being exposed to tourists, and tourism in a sense consequently informs identity according to outsiders' conceptions of indigenous culture. What this means is that people have to maintain an image that complies with the tourists' expectations–one of exotic, even primitive cultures largely unsullied by the outside world."

6. Demian (Forthcoming) has identified a similar connection between "loss" and "forget" among the Suau of Papua New Guinea. She claims that "People assert things that have been 'forgotten', such as songs, which old people will know and can sing. They mean that these songs are no longer part of everyday life because their place has been taken by new practices, new songs. So to claim that has been lost, and more significantly that it has been forgotten, is to say that this is knowledge which no longer connects up the world of persons and relations in a meaningful or effective way".

7. With reference to Indonesia, Dove (1988: 4) maintains that "allegiance to one of the world religions often has the quality of allegiance to a political party, with all the connections of expediency and impermanence that this implies." Conversely, the state nonrecognition of indigenous religious beliefs is seen by local communities as a discriminating factor against the employment of their schooled children in the Indonesia state, see Roy F. Ellen (1999: 150).

8. Not only in the Philippines but also in Papua New Guinea indigenous ideologies are being reassessed in a more positive light. Sillitoe (2000: 249) informs us that "an intriguing refiguring of relations is occurring between *kastom* and church. Whereas previously custom was associated by missionaries with savage pagans, ignorance and immoral behaviour, and church with civilised belief, enlightenment and property, today's Melanesians question this dichotomy. The positive reassessment of indigenous ideology and praxis with the emergence of *kastom* has given people the confidence to question trends in the wider world."

9. The resin is collected to be sold in order to raise cash (see Novellino 1999). Interviews with Batak elders reveal that the tapping of *Agathis philippinensis* (*bagtik*) is not a traditional activity. In the past, resin was gathered from tree branches in the high canopy or collected from the ground. The practice of tapping the tree trunks seems to have been introduced with the commercialization of forest products, probably after World War II.

10. Similarly, Kirsch (2001: 169) has argued that "the dynamics of memory and forgetting, the entropic tendencies of ritual knowledge, and the incompleteness of the intergenerational transmission of knowledge all pose questions about the possibility of loss. Yet loss may be integral to these systems in that it permits innovation and improvisation."

References

Appell, George N. 1988. "Costing Social Change." In *The Real and Imagined Role of Culture in Development: Case Studies from Indonesia,* ed. Michael R. Dove. Honolulu: University of Hawaii Press.

Asad, Talal. 1993. *Genealogies of Religions: Discipline and Reasons of Power in Christianity and Islam.* Baltimore and London: The Johns Hopkins University Press.

Benda-Beckman, K. von. 1984. *The Broken Stairways to Consensus: Village Justice and State Courts in Minangkabau.* Dordrecht: Foris Publications.

Demian, Melissa. Forthcoming. "Reflecting on Loss in Papua New Guinea." *Ethos* 71(4).

Dove, Michael R. 1988. "Introduction: Traditional Culture and Development in Contemporary Indonesia." In *The Real and Imagined Role of Culture in Development: Case Studies from Indonesia*, ed. Michael R. Dove. Honolulu: University of Hawaii Press.

Eder, James F. 1987. *On the Road to Tribal Extinction: Depopulation, Deculturation, and Maladaptation among the Batak of the Philippines.* Berkeley: University of California Press.

Ellen, Roy F. 1983. "Social Theory, Ethnography and the Understanding of Practical Islam in South-East Asia." In *Islam in South-East Asia*, ed. M. B. Hooker. Leiden: Brill.

___. 1997. "The Human Consequences of Deforestation in the Moluccas." In *Les Peuples des forêts tropicales. Systèmes traditionnels et développement rural des Afrique équatoriale, grande Amazonie et Asie du sud-est.* Civilisations vol. 44. no. 1-2. Brussels: Institut de Sociologie de l'Université Libre de Bruxelles.

___. 1999. "Forest Knowledge, Forest Transformation: Political Contingency, Historical Ecology and the Renegotiation of Nature in Central Seram." In *Transforming the*

Indonesian Uplands: Marginality, Power, and Production, ed. T. Murray Li. Amsterdam: Harwood Academic Publishers.

Foster, Robert J. 1992. "Commoditization and the Emergence of *Kastam* as a Cultural Category: A New Ireland Case in Comparative Perspective." *Oceania* 62: 284-94.

Geertz, Clifford. 1983. "Culture and Social Change: The Indonesian Case." *Man* 19: 511-32.

Hooker, M.B. 1974. "Adat and Islam in Malaya." *Bijdragen Tot de Taal-, Land- en Volkenkunde* 130: 69-90.

Jolly, Margaret. 1994. *Women of the Place: Kastom, Colonialism, and Gender in Vanuatu.* London: Routledge.

Josselin de Jong, J. P. B. de. 1960. "Islam versus adat in Negeri Sembilan (Malaya)." *Bijdragen Tot de Taal-, Land-, en Volkenkunde* 116 (1): 158-203.

Kirsch, Stuart. 2001. "Lost Worlds: Environmental Disaster, 'Culture Loss,' and the Law." *Current Anthropology* 42 (2): 167-98.

McDermott, Melanie H. 2000. *Boundaries and Pathways: Indigenous Identity, Ancestral Domain, and Forest Use in Palawan, the Philippines.* Dissertation, University of California, Berkeley.

Murray Li, Tania 2000. "Locating Indigenous Environmental Knowledge in Indonesia." In *Indigenous Environmental Knowledge and Its Transformations: Critical Anthropological Perspectives,* ed. Roy F. Ellen, Peter Parkes, and Alan Bicker. Amsterdam: Harwood Academic Publishers.

Novellino, Dario. 1997. *Social Capital in Theory and Practice.* Rome: Food and Agriculture Organization (FAO) of the United Nations.

____. 1999. "The Ominous Switch: From Indigenous Forest Management to Conservation–The Case of the Batak on Palawan Island, Philippines." In *Indigenous Peoples and Protected Areas in South and Southeast Asia,* ed. Marcus Colchester and Christian Erni. Document No. 97. Copenhagen: IWGIA.

____. 2000a. "Recognition of Ancestral Domain Claims on Palawan Island, the Philippines: Is There a Future?" *Land Reform: Land Settlement and Cooperatives* (1): 56-72.

____. 2000b. "Forest Conservation in Palawan." *Philippine Studies* 48: 347-72.

____. 2003a. "Miscommunication, Seduction, and Confession: Managing Local Knowledge in Participatory Development." In *Negotiating Local Knowledge,* ed. Johan Pottier, Alan Bicker, and Paul Sillitoe. London: Pluto Press.

____. 2003b. "Contrasting Landscapes, Conflicting Ontologies: Assessing Environmental Conservation on Palawan Island (the Philippines)." In *Ethnographies of Conservation: Environmentalism and the Distribution of Privilege,* ed. David G. Anderson and Eeva Berglund. London: Berghahn.

Olwig, Kenneth R. 1993. "Sexual Cosmology: Nation and Landscape at the Conceptual Interstices of Nature and Culture; or What Does Landscape Really Mean?" In *Landscape Politics and Perspectives,* ed. Barbara Bender. Providence and Oxford: Berg.

Peletz, Michael Gates. 1988. *A Share of the Harvest: Kinship, Property, and Social History Among the Malay of Rembau.* Berkeley: University of California Press.

Sillitoe, Paul. 2000. *Social Change in Melanesia: Development and History.* Cambridge: Cambridge University Press.

Spyer, Patricia. 2000. *The Memory of Trade. Modernity's Entanglements on an Eastern Indonesia Island.* Durham, NC: Duke University Press.

Stivens, Maila. 1996. *Matriliny and Modernity: Sexual Politics and Social Change in Rural Malaysia.* St. Leonards: Allen and Unwin.

Strathern, Marilyn. 1995. "The Nice Thing About Culture Is That Everyone Has It." In *Shifting Contexts: Transformations in Anthropological Knowledge*, ed. Marilyn Strathern. London: Routledge.

Tsing, Anna Lowenhaupt. 1999. "Becoming a Tribal Elder, and Other Green Development Fantasies." In *Transforming the Indonesian Uplands: Marginality, Power, and Production*, ed. T. Murray Li. Amsterdam: Harwood Academic Publishers.

Warren, Carol. 1993. *Adat and Dinas: Balinese Communities in the Indonesian State.* Kuala Lumpur: Oxford University Press.

Weiner, James F. 1999. "Culture in a Sealed Envelope: The Concealment of Australian Aboriginal Heritage and Tradition in the Hindmarsh Island Bridge Affair." *Journal of the Royal Anthropological Institute* 5: 193-210.

Zerner, Charles. 1995. "Through a Green Lens: The Construction of Customary Environmental Law and Community in Indonesia's Muluku Islands." *Law and Society Review* 28 (5): 1079-121.

Part Two

LOCAL PRACTICES: ADAPTIVE STRATEGIES AND STATE RESPONSES

Does Everyone Suffer Alike?

Race, Class, and Place in Hungarian Environmentalism

Krista Harper

"Everyone Is So Vulnerable"

While doing research on environmental politics and activism in Hungary, I interviewed an environmental lawyer in Budapest. We talked at length about how the sweeping legal and economic changes of the post-socialist transformation affected the environment. Referring to the broad array of environmental problems left in the wake of the transition from state socialism to the first stage of Eastern Europe's presumed inclusion in global capitalist markets, he said, "Environmental issues don't discriminate because everyone is so vulnerable."

The statement "everyone is so vulnerable" illustrates a widely held belief among Hungarian environmentalists. Activists frequently present the environment as a consensus politics: *everyone* benefits from clean air, water, and green spaces. The 1980s movement against the damming of the Danube River mobilized broad support, drawing on the significance of the river as a national symbol, historic site, natural monument, wildlife habitat, and drinking water source (Harper 2005). Danube activists presented environmentalism as a force for democratizing state socialist central planning, and in the late 1980s, the Danube movement was an important factor in the development of an organized political opposition. Its role in the expanding venues for public participation contributed to the democratic reputation of the environmental movement since 1989, reinforcing the notion that environmental politics cuts across class divisions and provides a "commons" for democratic participation because no one can ultimately escape the effects of environmental degradation of the natural commons.

Hungarian activists clung to the notion that environmental problems affect everyone equally, whether rich or poor, through the 1990s. After participating in environmental groups for about eight months, however, I began to pick up hints of emergent social inequalities in environmental debates. A few environmentalists broke with the trend of presenting environmental issues as divorced from socioeconomic and spatial inequalities and spoke out about the conditions of marginalized social groups. A variety of different activists articulated suspicions that specific groups suffered disproportionately from environmental pollution: low- and middle-income people living in cities, the rural poor, and Roma (Gypsy) communities in Hungary's post-socialist rust belt. These cases suggest that those who suffer most from the increasing socioeconomic disparities of the post-socialist period are more vulnerable to environmental degradation and illness as well: a post-socialist political ecology of human health. In the pages that follow, I focus on the intersections of environmental issues and Romani civil rights and public health. Activists' reflections on other forms of socioeconomic and spatial disparities inform my analysis of racial/ethnic disparities, however, and so I offer a brief discussion of class and urban/rural themes in Hungarian environmental discourses.

Class

Many activists described the environmental movement as a "middle-class movement," and socioeconomic class featured in environmental discourses with some frequency. I attended a meeting of the Clean Air Work Group in Budapest where one of the members, a middle-aged man in a dove-gray sweater, complained that public officials were not taking the problems of smog and public transportation seriously enough:

> We all know that the mayor and other important people in town value a green and healthy environment because they all move to expensive villas in the suburbs in the Buda Hills! Meanwhile, the rest of us poor folks have to live here in the city![1]

Another man stood up and announced to the group that a public official had bragged that the air of Budapest was better than that of Madrid or Paris:

> I don't know if up on the fourth floor of an office building on Nádor Street [a street near Parliament], they can't smell it, but those of us waiting for the bus down in the street know how bad the air in Pest gets! Do you remember a few years ago, it was February, and people were putting their scarves in front of their noses and mouths so they wouldn't have to breathe the smog? Maybe these officials don't come into contact with the air when they take their cars from the suburbs to their offices downtown, but we do![2]

Both activists use cars and suburban villas as symbols of the new, post-socialist elite to introduce issues of fairness and socioeconomic class into their discussion of air quality and traffic. Apparently, they suggest, one can buy cleaner air after all.

Place: Urban and Rural

The spatial dimension of urban/rural inequalities was another facet of the emerging political ecology of post-socialism to which activists drew my attention. A number of environmentalists advised me to pay attention to new environmental groups in the provinces (*vidék*) and not just to the Budapest environmentalist scene. In the 1990s, the number of new environmental groups outside Budapest grew substantially, and while Budapest groups are especially well situated to launch national campaigns and lobby with officials, environmental groups outside the capital city have a distinctive set of problems. Some environmentalists have maintained that these differences are rooted in the fact that outside Budapest, most of the country was rural and agricultural. While Budapest residents have relatively high incomes, with 40 percent of the population falling in the highest quintile, the nation's poor are concentrated in villages and rural areas (Szamuely 1996: 60).

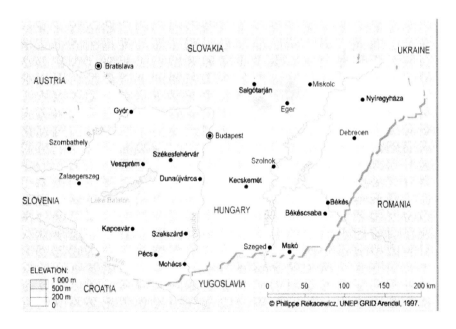

Figure 1. Hungary

The Budapest/provinces distinction parallels the differences between eastern and western Hungary (see Figure 1). Historically, the region west of the Danube (*Dunántúúli*) has been associated with Western Europe and the Hapsburg Empire, while the region east of the Tisza River (*Tiszántúli*) has been marginalized and associated with Hungary's "backwardness" (Sinkó 1989; Niedermüller

1989). During the state socialist era, the state attempted to develop the north-eastern counties by building up heavy industries, such as mining and steel production. With the political changes of 1989, many of the state-run factories were shut down, leaving the region with an unusually high unemployment rate and a major "brownfields" problem.

An activist from the Tree of Life Environmental Association in Eger showed me a map of Heves County's abandoned industrial sites and mines, many of which have left tailings. Slapping his hand to his face, he said, "We have to figure out which cases we can actually do something about. Whom can you sue over a formerly state-owned mine that has shut down, but hasn't been claimed by anyone?" In 1996, Tree of Life joined a number of other environmental groups in creating the Eastern Alliance, an environmental network focusing on the region's particular array of problems.

Ethnicity/Race: The Case of the Roma

Despite these signs of environmental activists' growing awareness of post-socialist socioeconomic inequalities, the issues of Hungary's large Roma minority remained invisible within environmentalist circles. The largest ethnic minority in Hungary, Roma constitute more than 5 percent of the nation's population (Havas et al. 1995). The transformation to a market economy has hurt Roma especially, as they make up two-thirds of the population living in poverty (Szalai 1999). At the same time, across Eastern Europe, Roma have served as scapegoats for the woes of the market economy, because popular stereotypes associate them with black marketeering, crime, and unemployment (Silverman 1995; Barány 1994). In the 1990s, Hungarian Roma faced violent attacks, police harassment, and local attempts to segregate schools, pubs, and restaurants (Furmann 1995, 1997).

Between 1989 and 2000, a small network of advocacy organizations and self-governments is attempting to address Roma civil rights and survival concerns, among them the Autonómia Foundation in Budapest. In the early 1990s, Autonómia was a part of Partnership for the Environment, an environmental foundation promoting grassroots groups. At that time, Autonómia worked on sustainable development grants for Roma communities. Autonómia formed a separate foundation in 1994 and continues to collaborate with Partnership for the Environment on sustainable development projects, but it focuses more on issues of unemployment, civil rights, and education. In 2004, the Coalition for Environmental Justice in Central and Eastern Europe (CEJ), an international network composed of lawyers, scholars, and activists from the Roma/human rights and environmental movements, was formed. The coalition has a number of participants from Hungary.

I asked a prominent environmental journalist, Klára Hoffmann, about the possibilities for environmentalist-Roma alliances. Hoffmann sighed and said,

"There are one or two people—not one or two groups, but one or two people—who are working on Gypsy issues …other than that, I have not heard of any Roma environmentalism." When I asked environmentalists why their groups did not deal with the problems of Roma communities, the most frequent response was that the main problems of Roma were poverty and access to education and that these were "social" issues, not environmental issues.

While several granting agencies provided some support for Roma sustainable development projects, grassroots environmental groups made almost no attempts to integrate Roma issues and activists into their campaigns. Like many other citizens in Hungary and east-central Europe, environmentalists expressed concern over the growing socioeconomic inequalities that have accompanied the transition to a market economy (Pine & Bridger 1998), but for the most part, this concern has not translated into environmental projects addressing poor and Roma communities.

From the beginning of my research with environmentalists, I had included a question about Roma communities in my regular interview schedule. For much of my fieldwork period, research participants appeared perplexed by the question, and their responses were cursory. I noted a marked change in participants' responses following a well-publicized case of lead poisoning in a Romani settlement. Research participants began to offer more elaborate answers about their views on the intersection of Roma concerns and environmental issues as they reflected on the case. I take up this case and the discussions that followed it later in this essay.

During my ethnographic fieldwork on the Hungarian environmentalist movement, I learned about several cases of pollution and toxic dumping in Roma villages and urban neighborhoods that resulted in illness and even death. I interviewed the director of a national-level Roma organization, the head of a foundation funding Roma initiatives, and two Roma journalists, and I participated in a conference of Roma civil rights activists and professionals.

The Roma activists and professionals whom I interviewed were familiar with the concept of environmental justice as an emerging component of minority civil rights struggles in the United States. Roma activists stated that they suspected that Roma and other low-income people were disproportionately exposed to toxic chemicals and air pollution. There was no research on ethnic or class disparities in environmental exposures, however, and few Romani organizers framed their group's issues in terms of environmental injustice.

Tragedy in Heves County

I first heard about the lead poisoning incident in Heves County from Aladár Horváth, head of the Foundation for Roma Civil Rights. We were sitting on the upper deck of a Danube ferry that had been chartered by the Green Alternative for a day of debates about the Danube River and activist networking. I had been conducting fieldwork for eleven months, and it was the first time I had ever encountered someone from the Roma civil rights movement at an environmentalist event. When we spoke in 1996, Horváth was interested in theories of ecoracism and environmental justice that were circulating in environmental studies and social movements in the United States. He expressed his sense that Roma settlements were disproportionately selected as dumping sites. Horváth had heard numerous stories about local governments that located sites for the disposal of dead animals near Roma neighborhoods, endangering public health. He was discouraged, however, by the lack of research on the environmental health conditions of Roma settlements. We discussed the connections between Roma issues and environmental issues, and the story Horváth told me that afternoon lingered in my mind for some time.

In 1995, tragedy struck the Roma settlement of the village of Heves. Men living in the village's Roma settlement found some discarded car batteries in nearby dumps and brought them home. They pried open the batteries in the backyard, removed the lead by melting it, and sold the salvaged metal to a local scrap dealer. As they processed the batteries by hand, lead got all over the yard, where people did housework and repairs and children played. One 15-month-old toddler died of lead poisoning, and two older children became seriously ill. Subsequent tests of the soil in the yard revealed that the quantity of lead was 270 times the public safety limit.

The father of the children, Sándor Kállai, was accused of serious bodily harm and homicide through negligence, as well as illegal battery demolition. The story appeared in newspapers and increased awareness of the plight of the poorest rural Gypsies. Following this episode, one Heves County Roma man won the lottery, and he donated some of the winnings to the poor squatters in Kállai's community. "I feel sorry for them because they have asked the mayor for help so many times," he told a reporter from the Roma magazine *Amaro Drom.* "I am also a Rom, and I know what it means to have nothing" (Doros 1996: 10).

Long before the court arrived at a verdict on Kállai, I heard several different analyses of the event from environmentalists and Roma activists. All of them drew attention to the special vulnerability of poor communities to economic and ecological exploitation. One environmental activist argued that a lack of access to civil organizations was a factor in the case. Environmentalists faulted the lack of regulation in the collection and processing of heavy metals.

Roma civil rights activists and environmentalists alike pointed to poverty and the dire unemployment of Gypsies in their analyses of the Heves lead poisoning incident. An Autonómia Foundation officer spoke with passion about the case:

> Some people might say that no sane, normal person would do what they did in Heves County. Now it is possible to say that these are not "normal" people—that they are

living at a subsistence level, and this is one way of subsisting, and they just didn't consider the fact that the children might get lead poisoning or how many centimeters into the ground they were polluting. This culture simply doesn't think about what will happen tomorrow. They are thinking, "If I bring this back to the house today, then maybe I can make some money out of it."

Her response calls attention to the fact that, for Heves's Roma residents, toxic waste salvaging is a means of livelihood as well as a threat to life.

Environmentalists related the Heves story to the deindustrialization of northeastern Hungary. Speaking of the incident, members of the Hatvan Environmentalist Association told me about a number of cases in Heves County where poor communities were approached by investors who wanted to build incinerators, metal processing plants, and toxic waste dumps, at the same time offering to build amenities such as sports centers to persuade town councils in poor villages. Referring to the lure of such investment schemes, one Hatvan environmentalist said, "It's like a drug!"

The Heves lead poisoning incident was just one of a number of toxic waste cases to hit Heves County since the transformation from state socialism. In 1991, an ad hoc, local environmental organization successfully opposed a company's plan to construct a battery processing plant in the village of Gyöngyösoroszi because the plant used an older technology that would melt down the metals, producing air pollution. The Heves lead poisoning in 1996 was followed by a number of toxic waste actions. During the summer of 1996, a mysterious oily sludge was found oozing from the ground at an army base west of Eger. Tree of Life alerted the press to the problem, and the Hungarian army cleaned up its mess. Meanwhile, the investors in the 1991 battery processing project reappeared in Apc, a village near Hatvan. Tree of Life and the Hatvan Environmentalist Association helped the projects' opponents negotiate with the Apc town council and demanded that an environmental impact report be prepared.

An activist from Eger's Tree of Life identified the Heves Roma community's lack of contact with social movement networks as a form of vulnerability. He pointed out that although there are forty-four environmental groups and numerous Roma civil organizations in the county, the village of Heves had neither an environmental nor Roma group. The presence of these types of civil organizations, he hypothesized, might have helped prevent the incident by offering resources, advocacy, or education. In the other toxic disputes of the summer of 1996, environmental organizations had contacted the press, garnered the support of sympathetic professionals, and confronted local governing bodies, the army bureaucracy, and corporations. He interpreted the absence of any civil organizations as a sign of the village's extreme marginalization.

Environmental groups in Heves County interpreted the lead poisoning as demonstrating the need for tighter regulation of the scrap metal business. In discussing the Heves case, the Eger activist immediately pointed out that there is no collection of used car batteries in Hungary, so batteries end up in landfills, and illegal dumping sites and near mechanics' shops. Although it is illegal to work with heavy metals without a permit, the scrap metal business is largely unmonitored. He faulted the scrap metal dealers, who knew that the metal was not being processed legally and therefore made lead salvaging an income opportunity for the very poor. He maintained that the national government should require large-battery collection and then recycle or dispose of batteries in an environmentally acceptable manner. In the long term, he thought that battery recycling should be considered part of the production process, rather than an unfair externality shouldered by taxpayers and the people who live near dumps.

In the court case, the prosecution charged that Kállai processed the batteries without a permit, willfully poisoning his children and the environment. The defense lawyers argued that most of the other men in the village also processed batteries without a permit because of their dire poverty and chronic unemployment. In May 1997, Kállai was acquitted, as reported by the Roma Press Center and the newspaper *Népszava*:

> The reason the court gave for the acquittal is that "it could not be expected from the accused, given his low intellectual capacity and poor education, to be careful to understand the consequences of the damages that can be done by battery demolition and lead." (*Népszava*, May 23, 1997)

In the final verdict, the judge emphasized the defendant's ignorance and lack of education as root causes of the tragedy.

Purity and Pollution Among Roma and Environmentalists

Upon further analysis, the lead poisoning incident exposes different cultural assumptions about pollution, waste, and production among middle-class environmentalists and poor Roma. Waste is a social relation like production, and it is strikingly important in the relationship between Roma and *gadje* (non-Roma). Historically, Hungarians have either romanticized Roma as carefree, creative spirits or denigrated them as profligates and social parasites. These twin views have contributed to their marginalization of Roma in feudal and capitalist peasant economies, in the state socialist collective economy, and in the post-socialist, deindustrializing economy.

Images of Gypsies: Noble Savages or Savage Profligates?

Environmental and development discourses across the world tend to portray marginalized and indigenous people in either of two ways: as noble savages or as environmental profligates. On the noble savage side, indigenous people are frequently cast as keepers of "primitive ecological wisdom" (Milton 1996: 109). Alternatively, the same people may be seen as environmental villains: slash-and-burn agriculturalists are blamed for deforestation and soil erosion, and hunting practices are interpreted as poaching (Ranger 1989).

Hungarian environmental activists avidly studied accounts of non-Western and indigenous peoples in their search for alternative, more environmentally sustainable cultures, but they rarely mentioned Roma in these terms. This may be due, in part, to the fact that Roma are not considered an indigenous people in Eastern Europe, having first arrived from the southeast in the fourteenth century. I did, however, meet one environmentalist who expressed his doubts about industrial development with reference to a romantic picture of Roma life:

> I've often felt their way of life is much easier on the environment. Perhaps the state shouldn't be teaching the Gypsies to live in apartment houses; maybe they should let them be, and allow them to teach other people about a life lived closer to nature and consuming less.

This environmentalist associates Roma with a preindustrial community, where traditional crafts and a less routinized daily schedule contribute to a more ecological, less alienated way of life and labor:

> They don't have as many machines—they always work with horses and hand tools …Their free way of life doesn't fit with the kind of life where you have to go to work every day in a factory, from morning till night.

Roma, as stereotypically imagined by this environmentalist, are marginalized in industrial society by virtue of their culture's very conviviality. His image of Roma life, however, bears little resemblance to the actual conditions of Hungarian Roma, the majority of whom integrated into industrial labor for decades and come from families that have been settled for generations.

This picture of Hungarian Roma conforms to romanticized images of "Gypsy life" that have prevailed throughout Central and Eastern Europe for at least two centuries (Lemon 2000; Pogány 2004). Hungarian Roma have frequently been stereotyped as spontaneous, carefree wanderers, despite the fact that only for the past century, only a small fraction of Hungarian Gypsies lived as nomads. In the mid-nineteenth century, the composer Franz Liszt wrote admiringly of Hungarian Gypsy musicians' unschooled, intuitive grasp of music and their spontaneous, improvisational creativity (Crowe 1996: 79). Like earlier non-Roma cultural producers, the environmentalist uses an idealized, preindustrial Roma community as a foil to alienating, bureaucratic, industrialized Hungarian society. Although presenting Roma as "closer to nature" serves to show how far industrial society has strayed, one must be cautious about the use of discourses on nature and biology with respect to Gypsies. The idea that Gypsies are "closer to nature" has historically reinforced racist beliefs that Gypsies are like wild animals and have a biological predilection for criminal behavior (Okely 1996; Bárány 1998). In the past decade, Hungarian nationalists reinvigorated these types of sociobiological claims about the true "nature" of Gypsies.

The view of Gypsy as noble savage posits Roma as a traditional culture endangered by the colonizing forces and development schemes of industrialism and state bureaucracies. More often, however, non-Roma Hungarians have characterized the Roma as dirty and profligate. While doing research in the environmental movement, I heard no activists openly castigate Gypsies as environmental profligates, but I came to see non-Roma beliefs about Roma pollution and profligacy as hidden cultural scripts that stymie cooperation between Roma and environmentalists.

Non-Romani Eastern Europeans frequently complain that Gypsies are dirty (Guglielmo 1993). The belief that Roma are unclean frequently is used to justify discriminatory practices. In 1997, the parents of non-Romani preschoolers requested that a village school hold segregated kindergarten graduation ceremonies on the grounds that Roma children were unsanitary (Kóczé 2000). Poor Roma living as squatters in abandoned houses on the periphery of rural Hungarian villages, like Heves, often lack plumbing, clean water, and electricity that most Hungarians take for granted (Guglielmo 1993).

The non-Romani symbolic association of Gypsies with pollution comes partly from the traditional Roma economic practices of scavenging and repairing discarded items. Anthropologist Michael Stewart (1993) analyzes how Gypsy scavenging has historically been vilified and contrasted with the hegemonic Hungarian work ethic. Prior to the socialist era, Roma could not own land, making their living instead from traditional crafts as musicians, tinkers, and blacksmiths, from scavenging, repairing, and reselling discarded objects, and from doing odd jobs for non-Roma peasants. Roma were excluded from the distribution of agricultural land after World War II, and in the 1950s, communist bureaucrats saw the Gypsy population as living outside class society. Only beginning in the 1960s did the state make any attempt to integrate them into wage labor. The Roma survival strategy of scavenging and repairing items for resale was called "usury" and discouraged by the authorities; yet, according to Stewart, scavenging and marketeering is one of the important ways Hungarian Roma have defined themselves in relation to non-Roma (Stewart 1993: 194; see also Okely 1996).

An administrator from Autonómia Foundation spoke about the relationship between Roma issues and environmentalism in the context of Gypsies and scavenging:

> If you go to a Gypsy settlement and look at classic slums, you can see that these clash in every way with the environmental notions that your usual middle-class person

supports … You can see, in these places, how there are bits and pieces of trash decorating a wall or patching the roof—there's a constant recycling and reproduction of waste, while in the proper middle class, there's a sharp distinction between what is garbage and what is not.

To illustrate her point, she showed me a documentary film, *Szemetlét* ("Life at the Dump"), which followed Roma scavengers from their rounds at the dump through their reuse of salvaged materials at home. Several men collected discarded bricks at the dump and used them to build a house. Another person brought home a long ribbon of red plastic with circles punched out in a regular pattern—surplus from a plant that packaged rounds of cheese. Back home, the red plastic was put to use marking the boundaries of the family vegetable plot.

The Autonómia administrator showed me this film because she wanted to counter the frequent negative portrayals of Roma scavenging and housekeeping practices as uniformly dirty and polluting. She pointed out that pollution is an interpretive category rich in cultural variation. Roma homes are maintained according to beliefs about pollution, but these beliefs differ from those of the non-Roma. "Usually," she said, "a Gypsy household is full of what you would call, from an environmental point of view, hazardous waste, which they collect planning somehow to use later." Roma scavengers show a great resourcefulness and make good use of objects carelessly discarded by industries and consumers.

Pollution and the Political Ecology of Work

Non-Roma fears of Roma pollution extend beyond cultural beliefs about scavenging into other areas of Roma employment. In the first decade of state socialism, Roma were officially considered outside the class system and excluded from wage labor in state-run industries (Stewart 1993: 187-88). In the 1960s, the state made sweeping efforts to assimilate Roma into waged labor, but for the most part, Roma laborers were placed in the most menial, low-paying positions available, and they were given the dirtiest and most physically demanding jobs in any industry (Crowe 1996). By the early 1980s, 75 percent of Roma workers were employed in industrial areas, overwhelmingly in unskilled jobs (Crowe 1996: 98).

Because of this ethnic division of labor under state socialism, Roma are associated with dirty work in heavy industries. The development of such industries was a high priority under state socialism because it symbolized modernization and industrial development (Burawoy 1985). Although Hungary is relatively mineral poor, compared to neighboring countries like Slovakia and Romania, the Hungarian state subsidized its steel and iron industries because of their symbolic importance. Since 1989, many of these mining and smelting factories have been shut down, creating an industrial rust belt in northeastern Hungary.

While the closing down of factories in the northeastern counties improved air quality and other environmental conditions, it has had devastating social consequences for the region. Unemployment skyrocketed, with unskilled, Roma workers especially hard hit. The administrator from Autonómia described this situation as an obstacle for Roma-environmentalist cooperation:

> Gypsies' issues and environmentalism clash with one another in Hungary today. That is to say, the jobs that Gypsies lost, the jobs that have been shut down, were in those industries that were horribly polluting to the environment. Take, for example, the demolition of the Ózd ironworks, or Miskolc's entire industry, which polluted to such an extent that it was scandalous, but nevertheless gave work to the Gypsies.

Heves County, the scene of the lead poisoning incident, is in northern Hungary, on the western margins of the rust belt. In the past, the county's economy was split between agriculture, tourism, and the mining of heavy metals, but since 1989, the heavy metal industry has been privatized, and without state subsidies, many of the mines have been closed. This industrial decline has left many Gypsy men in the region unemployed. The nationwide unemployment rate for Roma was 40-45 percent in 1993, almost three times the unemployment rate for the nation as a whole (Crowe 1996: 103). In towns in the northeastern rust belt, Roma unemployment was particularly high. In 1995, two-thirds of the national Roma population fell into the lowest quintile of income distribution (Andorka & Speder 1995, cited in Szamuely 1996: 60). Unemployment in the northeastern counties is much higher than the nationwide rate.

Rising social tensions in the northeastern counties came to a head in the early 1990s in a series of violent attacks on Roma by skinheads. In response to these attacks and the worsening conditions of Roma communities, Roma civil organizations in northeastern cities grew considerably. In 1993, several Roma groups organized a "March against Violence" in the city of Eger. Since unemployment, racial violence, and conditions of extreme poverty present poor Hungarian Roma communities with immediate challenges to survival, many Roma see environmentalism as a middle-class "luxury" cause with a distant future orientation or, at best, an abstract concept for another time. The administrator from Autonómia gave an example illustrating this difference in priorities:

> We tried to send a director of a local Gypsy organization to an environmental training session, but he felt like they just weren't concerned about the same issues that he was interested in. Gypsy groups care about survival in the short-term sense, while environmentalists are thinking much farther into the future.

While environmentalists often speak in terms of protecting the environment for future generations and dealing with the legacy of technological and industrial development, Roma civil rights activists see themselves as focusing on present conditions and the visible legacy of racism, poverty, and social exclusion.

The administrator from Autonómia traced the development of Hungarian environmentalism to the growth of the middle class under state socialism. She saw the environmental movement as a broad opportunity for opening up possibilities for activism and public debate:

> I think the environmental movement was launched out of the middle class, but then, that was the first place where it was possible to have civil initiatives. If there was any opportunity at all to oppose–to take a position in a civil matter where you could really intensify the debate–it was environmentalism. It was an area where Hungarian civil society could start to emerge at all, and that's important.

She continued:

> But the people who did this were able to do it because they had not only a secure material background but also a social background. They were educated people. Even if they didn't have jobs, they still had other forms of stability to fall back on.

In other words, she viewed Hungarian environmentalists' concern about unforeseen future consequences as belying a distinctively middle-class orientation to the future. This orientation is tied to the fact that environmentalists often homogenize their cause—"Everyone suffers when the environment suffers." Some environmentalists, however, are beginning to recognize the need for Roma civil rights and environmental projects that reflect their concerns.

Prospects for Change

In Heves County, not long after the lead poisoning incident, Hungary's first grassroots Roma environmental organization was formed, the Kerecsend Gypsy Environmentalist Association. One of the group's leaders approached Eger's environmental group, Tree of Life, with a local initiative to foster more cooperation between environmentalist and Roma groups. In their proposal, the Kerecsend Gypsy Environmentalist Association planned to clean up a creek bed where people illegally dumped trash. They requested funds to hire five workers to build a walking path and a small community center alongside the creek, and they proposed to restore the wetlands by planting reeds and reviving the basket-making craft. Since basketry was a traditional handicraft of Roma living in wetlands areas, this part of the project combined aspects of cultural preservation, ecological restoration, and sustainable development. Several of the staff members of Tree of Life and members of Roma community organizations assisted the Kerecsend group in the preparation of their grant proposal. The proposal was successful, and the Kerecsend group was able to obtain funding for several other small projects. The creek restoration proposal, which brought together activists from both environmental and Roma civil organizations, marked the first appearance of a self-designated Roma environmental group. The environmentalist-Roma activist collaboration at Kerecsend was short-lived, however, and few other environmental groups have forged such relationships with Roma organizations and communities in the past decade.

Hungarian environmental politics remain largely divorced from Roma activist organizations and their concerns, but there are emerging prospects for change within state institutions and social movements. Since joining the European Union, the Hungarian government has set up a Ministry of Equal Opportunities, and Hungarian Roma constitute one of the target populations for the government's antidiscrimination policies. As part of the state's push to mainstream Roma-related programs across the ministries, the Hungarian Environmental Ministry has taken three initial steps to encourage Roma participation in environmental decision-making, to address the environmental concerns of Roma communities, and to collect data on ethnicity and environmental conditions. The ministry set up a Roma reference bureau in 2004, headed by János Csonka, an environmental engineer of Romani descent. The Ministry's Green Source (*Zöld Forrás*) program funded a number of small environmental projects proposed by Roma community organizations between 2000 and 2005. Although some ministry officials expressed misgivings about the short-term nature of projects proposed by Roma organizations (such as neighborhood litter removal), the Green Source program has provided an important point of contact between the ministry and Roma citizens. Finally, the ministry commissioned researchers from the Debrecen University School of Public Health to map environmental conditions in Roma settlements in ten of Hungary's seventeen counties (Debrecen University School of Public Health 2004). The study surveyed Roma settlements' access to clean water, sewerage, gas, and electricity, as well as their placement relative to legal and illegal dumps, and it showed many cases in which Roma settlements had less access to a clean and healthy environment than their non-Roma neighbors.

Social movement and civil society actors are beginning to mobilize around issues of environmental justice in Hungary and in Central and Eastern Europe. Autonómia and Partnership for the Environment continue to provide small grants for sustainable development projects in Roma communities. Community organizations led by Roma leaders are beginning to make connections between livelihood issues, health, and environmental issues such as dumping and access to green space. These groups include the Sajó River Community Association of Sajószentpéter in northern Hungary and the Romarket Cooperative Association of Barcs in southwestern Hungary. At the national level, the Roma Civil Rights Association worked with community-level groups to fight air pollution from a power plant in Heves County in 2005. At the international level, the formation of the CEJ is bringing together a network of researchers, lawyers, and activists from environmental and Roma organizations. The group convened the first Transatlantic Initiative for Environmental Justice in October 2005, a meeting where North Americans and Central and Eastern Europeans came together to discuss the nature of environmental inequalities in their home countries and exchange strategies for promoting environmental justice. The actions of groups working at the local, national, and international levels are beginning to demand recognition of marginalized groups in environmental decision-making in Central and Eastern Europe.

Bodies and Landscapes at the Margins: Challenges for Sustainability

If they are committed to a truly democratic environmental politics, policy-makers and activists alike must reckon with the social and environmental marginalization of Roma people as they address environmental problems and sustainable development in Central and Eastern Europe. Environmental scholars and activists in the United States are now bringing greater attention to ethnic and class inequalities and sustainable development (Wenz 1994; Bullard 1994). With the complex reorganization of post-socialist property relations and socioeconomic class, Hungary's environmental and social movements stand to gain from taking inequalities and different experiences of environmental degradation into account (Pavlínek & Pickles 2000).

Environmental historian Robert Gottlieb (1993) suggests that paying attention to the intersections of public health and environmentalism provides a crucial link for environmental justice: how the marginalization of places relates to the marginalization of bodies. The marginalization of places and bodies is further complicated by the historically specific conditions of late and post-socialist transformations. In his research on rangeland privatization in Inner Mongolia, anthropologist Dee Mack Williams (1997) examines how changes in the Chinese state's orientation toward the global economy transform the physical landscape and the health and life chances of people living in marginal lands. Williams calls for anthropologists to pay attention to the ecological and corporeal ramifications of globalization.

By the mid-1990s in Hungary, environmentalists began to see changes in the landscape wrought by post-socialist privatization, part of a broader trend across the entire post-socialist region (Pavlínek & Pickles 2000). Urban environmentalists also pointed to the bodies of children—suffering from chronic respiratory ailments—as corporeal evidence of the air pollution brought by an increasingly automobile-oriented consumer culture. Within an environment increasingly marked by socioeconomic inequalities, Gypsies' settlements and bodies are especially vulnerable to environmental risks.

While everyone may suffer from environmental degradation, it seems equally clear that we do not all suffer alike. Recognition of difference, paradoxically, might lead to more inclusive environmental politics—one that builds upon the democratic legacy of the Hungarian environmental movement. The Autónómia administrator pointed to the possibility of redefining what is included in "environment" and "sustainable development." "If we take a broader definition of sustainable development, making it socially sustainable also, then I think it will be a more inclusive project," she told me. "If it's not just about whether we ruin the physical environment for the next generation, but also about making the social environment more livable, then that is a very important step." An environmental movement capable of recognizing class and racial inequalities could be better equipped to confront the broader ecological and public health consequences of post-socialist industrial decline and integration into the European Union.

Paying closer attention to the intersections of social and economic marginalization, health, and place presents the possibility of bridging the gap between environmental and Roma civil organizations in Hungary, pointing to their shared interest in democratic participation, equitable urban planning, and sustainable development. Doing so would expand environmental mobilization to include emergent issues of environmental justice among Hungarian Roma and set an example for the inclusion of marginalized groups in environmental decisions across Europe.

Acknowledgments

Field research related to this project was supported by an IREX Individual Advanced Research Opportunity grant (1995), a Fulbright IIE Graduate Research Fellowship (1996), and travel and research grants from the Political Economy Research Institute (PERI) at the University of Massachusetts (2000) and the Smith College Dean of Faculty (2002). The author wishes to thank Alaina Lemon, Carl Maida, S. Ravi Rajan, and James K. Boyce for their comments on this paper.

Notes

1. Author's translation (from Hungarian) of speech recorded in fieldnotes from a public meeting of the Clean Air Action Group, Spring 1996.
2. Author's translation of speech recorded in fieldnotes from a public meeting of the Clean Air Action Group, Spring 1996.

References

Bárány, Zoltán. 1994. "Nobody's Children: The Resurgence of Nationalism and the Status of Gypsies in Post-Communist Eastern Europe." In *East-Central Europe in the 1990s,* ed. Joan Serafin. San Francisco: Westview.

___. 1998. "Ethnic Mobilization and the State: The Roma in Eastern Europe." *Ethnic and Racial Studies* 21 (2): 308-27.

Bullard, Robert. 1994. "Environmental Racism and the Environmental Justice Movement." In *Key Concepts in Critical Theory: Ecology,* ed. Carolyn Merchant. Atlantic Highlands, NJ: Humanities Press.

Burawoy, Michael. 1985. *The Politics of Production: Factory Regimes Under Capitalism and Socialism.* New York: Verso.

Crowe, David. 1996. *A History of the Gypsies of Eastern Europe and Russia.* New York: St. Martin's Griffin.

Debrecen University School of Public Health. 2004. *Telepek és Telepszerü Lakóhelyek Felmérése* ("Survey of Settlements and Settlement-type Residences"). Debrecen, Hungary: Debrecen University School of Public Health.

Doros, Judit. 1996. *"Ajándék Százezer Forintért: Nyert a Lottón, Élelmiszert Vett a Rászorulóknak"* ("100,000 Forint Gift: He Won the Lottery, Bought Groceries for the Needy"). *Amaro Drom* 6 (2): 10.

Furmann, Imre. 1995. *The White Booklet: Based on the Cases of the Otherness Foundation's (Másság Alapítvány) Legal Defence Bureau for National and Ethnic Minorities.* Budapest: Másság Alapítvány.

___. 1997. *Annual Report.* Budapest: Másság Alapítvány.

Gottlieb, Robert. 1993. "Reconstructing Environmentalism: Complex Movements, Diverse Roots." *Environmental History Review* 17 (4): 1-20.

Guglielmo, Rachel. 1993. *Milyen Út Var Rájuk?/The Gypsy Road.* Nagykanizsa, Hungary: The U.S. Peace Corps and Amalipe Cultural Preservation Society.

Harper, Krista. 2005. "'Wild Capitalism' and 'Ecocolonialism': A Tale of Two Rivers." *American Anthropologist* 105 (2): 221-33.

Havas, G., *et al.* 1995. "The Statistics of Deprivation." *Hungarian Quarterly* 36: 67-80.

Kóczé, Angéla. 2000. "Romani Children and the Right to Education in Central and Eastern Europe." *Roma Rights* 3. http://www.errc.org/cikk.php?cikk=1008.

Lemon, Alaina. 2000. *Between Two Fires: Gypsy Performance and Romani Memory from Pushkin to Postsocialism.* Durham, NC: Duke University Press.

Milton, Kay. 1996. *Environmentalism and Cultural Theory: Exploring the Role of Anthropology in Environmental Discourse.* New York: Routledge.

Niedermüller, Péter. 1989. "National Culture: Symbols and Reality." *Ethnologia Europaea* 19 (1): 47-56.

Okely, Judith. 1996. *Own or Other Culture.* New York: Routledge.

Pavlínek, Petr, and John Pickles. 2000. *Environmental Transitions: Transformation and Ecological Defense in Central and Eastern Europe.* New York: Routledge.

Pine, Frances, and Sue Bridger. 1998. "Introduction." In *Surviving Post-Socialism: Local Strategies and Regional Responses in Eastern Europe and the Former Soviet Union,* ed. Sue Bridger and Frances Pine. New York: Routledge.

Pogány, István. 2004. *Roma Café: Human Rights and the Plight of the Romani People.* London: Pluto Press.

Ranger, Terence. 1989. "Whose Heritage? The Case of the Matobo National Park." *Journal of Southern African Studies* 15 (2): 217-49.

Silverman, Carol. 1995. "Roma of Shuto Orizari, Macedonia: Class, Politics, and Community." In *East European Communities: The Struggle for Balance in Turbulent Times,* ed. David Kideckel. San Francisco: Westview.

Sinkó, Katalin. 1989. "Árpád versus Saint István: Competing Heroes and Competing Interests in the Figurative Representation of Hungarian History." *Ethnologia Europaea* 19 (1): 67-84.

Stewart, Michael. 1993. "Gypsies, the Work Ethic, and Hungarian Socialism." In *Socialism: Ideals, Ideologies, and Local Practice,* ed. C. M. Hann. New York: Routledge.

Szalai, Julia. 1999. "Recent Trends in Poverty in Hungary." In *Poverty in Transition and Transition in Poverty,* ed. Yogesh Atal. New York: Berghahn.

Szamuely, László. 1996. "The Social Costs of Transformation in Central and Eastern Europe." *Hungarian Quarterly* 37 (144): 54-69.

Wenz, Peter. 1994. "The Importance of Environmental Justice." In *Key Concepts in Critical Theory: Ecology,* ed. Carolyn Merchant. Atlantic Highlands, NJ: Humanities Press.

Williams, Dee Mack. 1997. "Grazing the Body: Violations of Land and Limb in Inner Mongolia." *American Ethnologist* 24 (4): 763-85.

Attachment Sustains
The Glue of Prepared Food

Deborah Pellow

The geographer Mabogunje (1990: 128) has observed that one of the early problematics of African urbanization focused on "demographic development and the emergent social differentiation among urban Africans." This raised interest in so-called segregational issues, such as the ethnic segregation that grew out of migration histories and that was kept alive by colonial policies. More recently, writers like Hannerz (1987, 1992; see also Mintz & Price 1985) have moved on to consider the manner in which separate cultures come together and hybridize, suggesting that "this world of movement and mixture [as] a world in creolisation" (Hannerz 1987: 551) is a promising metaphor for the African urban milieu as part of the globalizing world.

> At one end of the creolizing continuum there is the culture of the center, with its greater prestige … at the other end are the cultural forms of the farthest periphery, probably in greater parochial variety. (Hannerz 1992: 264)

Groups organize themselves along this continuum, "mixing, observing each other, and commenting on each other" (ibid.: 265). There are asymmetries of cultural flow, with different points of origin and reach.

This chapter visits a community in Accra, Ghana, that was founded at the beginning of the twentieth century for migrant Hausa from northern Nigeria.[1] These people brought with them a cornucopia of customs and traditions, including cuisine. I focus upon the importance of prepared food as something that matters (Miller 1998), for it is in part through its consumption that this particular sociospatial community is sustained. It helps perpetuate community identity and vitality, as a place where residents can buy food for basic daily sustenance, special occasions, and cultural nostalgia, and where nonresidents can procure specialties not sold elsewhere. Food is similar to all culturally defined material substances in that it is used to create and maintain social relationships, to solidify membership, to define groups. "Like languages and all other socially acquired group habits, food systems dramatically demonstrate the infraspecific variability of humankind ... food preferences are close to the center of [a group's] self-definition" (Mintz 1985: 3; see also Mintz & DuBois 2002). Food is like a language; through it people articulate and acknowledge their distinctiveness (Counihan 1999). Specific foods carry specific meanings, and people's consumption of those foods signifies those meanings, for example, of culture, sex, or status.

In this local community, a concentration of people engages in complex networks of social relations (Leeds 1973; Pellow 2002). This paper explores one way in which practices of food consumption help sustain these relations and the community that they constitute. As Maida has observed in his introduction to this book, for sustainability to endure, social, economic, and environmental factors within human communities must be viewed interactively. Different communities have different needs and quality of life concerns, and communities are sustained when local practices allow or enable the satisfaction of those needs and concerns.

This essay illustrates the significance of prepared food specialties and their spatial arrangement to the community and the practice of everyday life, to the definition and sustenance of Sabon Zongo and the reinforcement of peoples' attachment to that place. The prepared food that is sold and consumed occupies various spots along the continuum of creolization. This paper's concern is with those specific to Hausa cuisine, but also significant are the foods characteristic of southern Ghana and Accra in particular. Both the Hausa and the Ghanaian eatables are made available at particular sites. This accords certain places a kind of power, bringing together what Agnew and Duncan (1989: 1) refer to as "geographical and sociological 'imaginations.'" The current study attends to two such places in Sabon Zongo as sites of consumption: particular street trading locales, where Hausa specialties are sold, and the community's Night Market, where prepared food traders sell southern Ghanaian food. The two locales and their products exemplify the varying intersections of cultural streams and creolization. While food habits are integral to cultural behavior and are often identified with specific groups, acculturation does allow people to adopt or adapt "foreign" foods (Fieldhouse 1986).

The Hausa in Accra

Accra is made up of diverse groups, one of which is the Muslim community (see Pellow 1985). The Muslims' collectivity was established by migrant ethnic groups, primarily from Nigeria. As Dretke (1968: 10) observed in his thesis on

Accra's Muslim community, "Wherever trade went, there was a Muslim trader." Since the late fifteenth century, Muslim traders have been present in the Gold Coast (pre-Independence Ghana), and the West African trade in kola nuts, spurred on by Usman dan Fodio's 1804 *jihad*[2] resulted in Muslim traders traveling to the northern part of the Gold Coast, particularly Dagomba and eastern Gonja, to buy the nuts. The significant nineteenth-century market of Salaga in eastern Gonja declined after 1874, and many Muslim traders relocated to Accra. In addition, Muslims recruited into the Gold Coast Hausa Constabulary, organized by the British in 1870, were brought to Accra in the late 1880s (see Figure 1).

WEST AFRICA

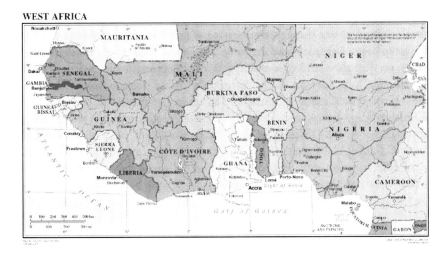

Figure 1. West Africa

Accra's population of Muslims in 1891 was about 8 percent of the total population of 19,999. They lived together in a *zongo,* a stranger quarter. It was a diasporic community, whose major ethnic groups included Yoruba, Hausa, Kanuri, Fulani, Nupe, and Wangara. While the Yoruba were probably most numerous, the enclave has always had a conspicuously Hausa influence (Pellow 1985). The Hausa gained a disproportionate influence on the others in religion and learning, economic affairs, roles and offices, dress, and language. Accra became a magnet for Hausa for several reasons. The Hausa had been active in trade throughout West Africa as early as the sixteenth century . Especially in kola and cattle, trade was an important source of wealth in Accra, as elsewhere in the Hausa diaspora (Cohen 1965), and was thus a basis for patronage. In addition to coming as traders and soldiers, they came also as teachers [*malam, malamai* pl.]. They came from the northern Nigerian Hausa "homeland," from such cities as Katsina, Kaduna, and Kano. Many brought wives and families. Rather than integrate with the growing cosmopolitan community of Accra, the Hausa stayed within the strangers' domain and worked as individual entrepreneurs.

The first of the Hausa to arrive in Accra is said to be Malam Idris Bako Nenu. He came from Katsina around 1860 (Pellow 2002) and received from the Ga chief a place to live among the other Muslims in the old Ga core near Jamestown Fort in Central Accra, an area that came to be known as Zongo Lane. Malam Nenu became a classic Hausa patron, a *mai gida,* taking in and helping new Hausa migrants. His son Malam Bako inherited his role of patron in 1899.

Sabon Zongo, the New Zongo

While Accra was already populated when the British arrived in the mid-nineteenth century, it was through the impetus of the colonial government that the city really began to expand and diversify. The spatial imprint of European imperialism on the urban landscape and its system of organization is evident in Accra (Pellow 2002). At the turn of the century, the city of Accra consisted of the so-called Central area, along the Atlantic Ocean, in the shadow of the three forts (Christiansborg, James, and Ussher), and the Administration's settlement in Christiansborg (Osu) three miles away (Lisowska 1984). Individual neighborhoods did not initially separate the indigenous people from the Europeans, but social differentiation evolved as did neighborhoods with distinct locations. The British constructed areas based on differences among the indigenous people, cordoning off sections for various ethnic groups and reinforcing the social and spatial compartmentalization of the town (Brand 1972). They also separated themselves off from the Africans, building a racecourse as a *cordon sanitaire* (Larbi 1996).

In 1908, the bubonic plague hit the old core of Accra; in this densely populated area, many died. To deal with sanitation in part by relieving the crowding, the governor initiated a program of town planning, renewal, and clearance, which included newly zoned town sites. One initiative was to tear down dwellings in Central Accra where crowding was extreme. This was coincident with the Hausa becoming more numerous, as well as with fighting within the growing Muslim community. The patron Malam Bako devised a plan for a new Hausa settlement, which served the British agenda of social and spatial compartmentalization. He established Sabon Zongo around 1910.

Accra's Sabon Zongo ("new *zongo*") was thus founded in the first decade of the twentieth century in part as a refuge for Hausa who had been living in the old *zongo* in the city's core, who wanted to get away from the squabbles and practice their Hausa culture. Initially Sabon Zongo's location was outside of the city limits, about two miles northwest of Accra's original nucleus, in what was then considered "bush." To get there, one had to cross the Korle lagoon by ferry. Over the years, the city has expanded and has grown around and enclosed Sabon Zongo. It is bounded on the north by a major road, used by commuters from suburban and peri-urban areas traveling to downtown Accra. In contrast to the different façades that the city of Accra presents, Sabon Zongo's population is heavily "northern";[3] while the population has diversified considerably, a heavy Hausa influence continues, intertwining elements of Islam and Hausa custom and practice. This includes the types of foods prepared and sold and the manner of selling.

Commerce in the Informal Sector

Accra, the capital of Ghana, has been the country's primate city since the British moved their administrative headquarters there in 1877. The city took shape under British site planning, and like many colonial cities, its spatial layout distinguished different neighborhoods based upon racial and ethnic differences (Mabogunje 1990: 138). These areas included the old Ga core, the European section, and the stranger area (in West Africa, the Muslim *zongo*).

On the one hand, Accra is a cosmopolitan administrative center, with an international airport, several four-star hotels, and an outward-looking Western ethos. Following the Structural Adjustment Program of the early 1980s, Ghana positioned itself as welcoming to foreign companies, and 95 percent entering settled in Accra. At the turn of the twenty-first century, there were 655 foreign companies from 80 countries operating in Accra (Grant 2002).

On the other hand, in urban Ghana in 1990, 60 to 85 percent of all employed people worked in the informal sector (Boeh-Ocansey 1997). While the term "informal sector" is used globally, it was coined by Keith Hart (1973) in his study of urban income opportunities and employment in Ghana. His basic distinction between formal and informal income opportunities was "that between wage-earning and self-employment," which he identifies with the sub-proletariat of the slum: "denied success by the formal opportunity structure ... members of the urban sub-proletariat seek informal means of increasing their incomes ... in this world of economic activities outside the organized labour force" (Hart 1973: 67-68). These informal jobs may be legitimate, for example, self-employed farmers and artisans, street hawkers, carriers, and petty service providers, or illegitimate, including hustlers, prostitutes, and gamblers.

The commerce that is responsible for Accra's growth and prosperity, and may be its single most conspicuous social feature, is informal sector trade (Dakubu 1997). Trading is the largest occupational category in Ghana, and 32 percent of Accra's employed are traders. Most have little or no education, and earn very little, and their businesses are not registered. "[T]hese are the producers and suppliers of most of the material requirements and services needed by the rest of Ghanaian society" (ibid.: 13).

The importance of trade to Accra goes back several hundred years. As early as the mid-seventeenth century, Accra was "the major center for the gold trade on the West African coast" (Robertson 1984: 29). In the mid-eighteenth century, Accra was a center for trade from Asante, the north, and the east. And in the nineteenth century, the principal commercial items that took over following the abolition of the slave trade were ivory and gold dust in exchange for gunpowder, rum, and cloth. In the second half of the nineteenth century, as many local Ga men shifted their occupations from farming and fishing to skilled labor, there was a coincident increase in educational opportunities and positions in the civil service for them (Robertson 1984).

Throughout West Africa, as urbanization flourished under colonialism, so did market trade. And as in Accra, with men taking specialized jobs, the market became the province of women. Urban women's primary trade has been in foodstuffs, and women's participation expanded along with the markets. According to Robertson (1984: 77), "approximately 70 to 75% of Accra Ga women were traders in the twentieth century." While the majority of the market women today are not big earners, the market in Accra is still one of "the last realms of influence left to women" (Robertson 1983: 476). And as observed thirty years ago, "'if tropical markets for basic foodstuffs worked less well, we should probably know a great deal more about them'" (Jones cited in Mabogunje 1990: 165).

On June 4, 1979, Flight Lieutenant Jerry John Rawlings, a young officer in Ghana's air force, led a successful coup d'etat, overturning the Supreme Military Council II, headed by General Fred Akuffo. During the preceding nine years, following the coup that had toppled the government of Prime Minister KA Busia, Ghana was ruled by a series of corrupt generals. This period of personal greed produced economic crisis. During the summer of 1979, as Rawlings's government imposed price controls and goods disappeared from circulation, the women traders were blamed for Ghana's economic woes. As a "cure," Makola Market, the primary wholesale and retail market in Accra, was razed to the ground by Rawlings's military regime (Robertson 1983). The fury with which many regarded the traders (and the conviction that the traders' success was based in occult practices) is evident in a letter I received a couple of months later:

> It is because the market was the centre of ... economic sabotage. After the mowing down of the Makola Market, so many things came to light. Could you imagine a two and a half (2½) year old baby (dead, with one big eye cirlating [sic] like a red light on top of an ambulance car) confined in a coffin and vomiting various sums of our local currency. Secondly a cat dressed in suit with spectacles on. Pythons vomiting money were uncountable ... In the course of the razing down of the market a woman came weeping and begged the soldiers to allow her remove her important personal thing from where her stall was located. The soldiers allowed her and the woman walked straight to where her stall was, knelt down and recited some incantations, and to the astonishment of the whole crowd, a big black snake sneaked into the woman's bossom [sic] and she was escorted. (Letter from E. Anyan, October 23,1979)

While the stories were the stuff of mythology, the letter testifies to the extraordinary degree of power attributed to the market women.

Trading still carries the greatest financial potential for the untrained or illiterate woman. Most traders operate downtown in Accra's central business district, bounded by the Ring Road and the coast. The core of this district is an outgrowth of the traditional Ga section in downtown Accra, home of several of Accra's famous markets–Makola, Salaga, and Timber Market.

But every neighborhood seems to have its local market, such as those at Adabraka and Kaneshie. While women may prefer to shop at the downtown markets, believing they have the best variety, the freshest produce, and the best prices, a local market such as Sabon Zongo's Freedom Market makes the daily life of a woman who resides nearby far easier. One of the women's primary domestic roles is to provide meals. If a *zongo* woman has the time and money to cook, but not enough time to go downtown to Makola or Salaga Market, she can run over to Freedom Market, which carries provisions, fresh foods, and cooking needs, as well as notions of various sorts. It is not particularly well known and is somewhat dilapidated, but given its location on the southern edge of the community, women shop there out of convenience.

Sabon Zongo has two other levels of market trade. The Night Market is located at the heart of the community; here women trade in local (Accra) prepared foods. And there is also the street selling of prepared foods, which to outsiders may seem spontaneous but in fact, as elsewhere in Accra, is actually quite formal in terms of the product sold and where and when it is sold. The foods sold fall along Hannerz's continuum of creolization. It may be the street trade in food that helps particularly to define and sustain Sabon Zongo as a unique community.

Creolized Food

Prepared food is a real draw to Sabon Zongo and represents much of its life blood: Hausa and other Muslims in and about Accra, as well as other Ghanaians, come for particular foods, knowing when and where they are sold. The foods lie along the continuum of creolization: through networks of interrelationships, people develop awareness of and familiarity with the cultural forms (cuisines) of others, creating a new system of meaning (Hannerz 1987). Thus, the Hausa who settled here brought along their food tradition, and the uniquely Hausa foods continue to be available. Sabon Zongo is a bonafide community in Accra, and there are Accra people (for example, Ga) who live here or come daily to sell their foods. The purveyors of Western imports provide adaptations of each and the fusions of all. When I was doing research in the late 1990s, the largest repertoire of Hausa foods in Accra was primarily available in this single community. In fact, prepared Hausa chop is so readily available in Sabon Zongo that a Hausa woman who lives elsewhere in the city claimed that girls brought up in Sabon Zongo do not learn how to cook. With the exception of meat, as elsewhere in Accra, which is men's work, prepared food selling is woman's work. As a prepared food trader said to me years ago, "men cannot feel to do it and many women cannot feel to do it." She was using as an example *kenkey*, Accra's maize staple traditionally prepared by boiling balls of mixed portions of fermented cooked maize meal and raw maize dough wrapped in cornhusks. And she continued:

> As for *kenkey*, if you want to make it, you will waste your time many hours, that's why men can't feel to do it … if you saw somebody was selling *kenkey*, she thinks her mother or her grandmother learning the *kenkey* before she starts. When you want to start the *kenkey* you will put the maize in the water. When you put it in the water, you should wash the pan first. Then take it to the machine and they will grind it. When they bring it back again, then you can't put your hand in it. You will wash it, you can't put your hand into it … Because *kenkey* will take more than six hours, and office job you can do in two hours. (Interview September 13, 1971)

The sellers of all types of prepared food have their respective niches, both spatial and temporal (see Table 1): the particular time of the day, depending upon what they prepare, and the location, which remains the same. While the daily minimum wage as of April 2004 is $1.22 (Radio Station JOY FM March 21, 2004), 50 percent of the population in Ghana lives on $1/day or less. Thus, street food is a necessity: it is less expensive to buy a meal on the street than to prepare it. In preparing soup, for family or for sale, one needs the fuel, the meat, the spices, and so on. In purchasing a bowl of soup, one chooses how much of the starch to include and may go without meat, thereby saving money. Selling particular foods at particular times helps reaffirm the social group(s). "[B]eloved foods are the taste of our lives, a steady source of comfort and succour in hard times," and to immigrants, foods "come to represent the home life they are denied" (MacClancey 1992: 18). Foods organize peoples' days, weeks, and months; *masa,* a corn-based cake that is fried in a steel form, is weekly holiday fare and sold Thursday night and Friday morning.

Morning foods are sold as early as 5:30 A.M. until they run out. *Wake,* beans, is sold with *gari,* coarsely ground cassava flour, can be eaten with snack foods such as fried plantain, and may be combined with rice and *taliya,* homemade pasta. The type of starch (e.g., rice, pasta, *banku*) is indicative of what food form is eaten–for example, rice or pasta with stew, *banku* (polenta-like maize) with okra stew. Bread can be bought by the loaf or by the slice. Those selling slices slather them with margarine from a tin. *Funkaso (pinkaso* in Ghana Hausa) are wheaten cakes and like *k'osai,* bean cakes, may be eaten together with *koko,* Ghana porridge. The prepared food sellers will supply utensils or one may bring one's own and "takeaway." *Wagashi* (fried curd) is sold by Hausa

Table 1. Prepared Food Street Trade

Hausa foods	Ghanaian foods

Morning

Hausa foods	Ghanaian foods
1. Wake	A. Banku
2. Rice	B. Koko (Ewe or Hausa porridge)
3. Taliya	C. Fried plantain
4. Wagashi (from Nupe)	D. Bread
5. Pinkaso (funkaso)	
6. Masa	
7. K'osai	

Midday/afternoon

Hausa foods	Ghanaian foods
8. Tuwo	E. Corn
9. K'afa	F. Kenkey
7. K'osei/tobani (bean cakes mixed with flour)	G. Cocoa yam (gwaza)
2. Rice	
10. Fura/nakiri (Mali and Wangara fura)	

Late afternoon/Evening
Night Market

Hausa foods	Ghanaian foods
2. Rice	F. Kenkey
	H. Kelewele
	J. Fried chicken

Street

Hausa foods	Ghanaian foods
8. Tuwo	F. Kenkey
2. Rice	J. Fried Chicken
11. Kebabs	A. Banku
3. Taliya	K. Fried yam
6. Masa (Thursday evening)	L. Omelet–with tea
12. Lamb/balangu	E. Corn
13. Lamoji (a drink)	H. Kelewele
14. Fresh milk, yogurt (Fulani)	M. Gari
1. Wake	N. Tea–with bread or omelet
5. Pinkaso	G. Cocoa yam (gwaza)
	B. Koko (Ewe or Hausa porridge)
	P. "Yo ke gari" ("red, red," plantain, beans, and palm oil)

women as a snack (it is too expensive for most to make a meal out of), but according to my sources, the food is more common among the Nupe than the Hausa in northern Nigeria.

Midday/early afternoon Hausa foods are different than those eaten in the early morning, and in Sabon Zongo, unlike other parts of Accra, the Ghanaian foods are different as well. For example, *kenkey,* the Ga staple, is eaten morning, noon, or night; in Sabon Zongo, it is not sold in the morning. *Fura* is a Hausa porridge; it is made from balls of millet that have been roasted and pounded with spices, then added to either milk or yogurt and sweetened with sugar. To find *fura,* one needs local knowledge, because those who sell it are not out on the street. The seller may supply a calabash bowl and the mixings, or one may take the millet ball home to prepare the dish there. *Tuwo,* the classic northern Nigerian staple, in Sabon Zongo is made from rice or cassava flour; it is cooked until the grains disappear, formed into balls, and served with soup. Soups are specifically Hausa and include *kuka* (leaves from the baobab tree), *k'ubewa* (okra, though different than southern Ghanaian okra soup), and *taushe* (no translation). Rice is served with stew. *K'afa,* a food of the Togolese Kotokoli, is an opaque and glutinous starch made from sieved maize; it is sold wrapped in a green leaf and is often eaten with peanuts. *Kenkey* may be sold alone or with fried fish and ground fresh pepper. Corn in season is grilled on the fire or boiled and often sold with chunks of fresh coconut. And cocoa yam is deep fried, like french fries, and eaten as a snack.

The Night Market gets under way after dark, around 6:15 P.M., and each seller has a kerosene candle on her table. The only Hausa chop sold at the Night Market is rice and stew. The substantial food is *kenkey* and fried chicken. *Kelewele* is a favorite snack throughout Accra: very ripe sliced plantain, mixed with ground pepper, ginger, and shallots and deep-fried. The evening prepared food street sellers begin to take up their posts on the street at 4:30 or 5 P.M. Among the first are the Fulani, who bring the fresh milk and fresh yogurt; every evening they sell by the corner barber, a gathering spot for men throughout the day. Many of the same foods–Hausa and Ghanaian–are sold in the evening as at other times of the day, with the exception of the kebabs and grilled lamb, tea, and omelets. The meat is prepared and sold by men only. They spend hours cutting up the animals and threading the meat on skewers (*cincinga,* kebabs) or laying slabs of lamb on a grill (*balangu*), which are sold in chunks with spices sprinkled on top. The meat is an attraction to outsiders–Hausa and non-Hausa. For many in the *zongo,* a taste is all that is affordable. Omelets are prepared to order and eaten by the side of the road.

By looking at Figures 2 and 3, one immediately notices patterns: nowhere or at any time is only Hausa or only Ghanaian food sold. The prepared food sold on the street is sold primarily in the old core near the Night Market and its periphery, along Korle Bu Street. This is a densely populated part of the

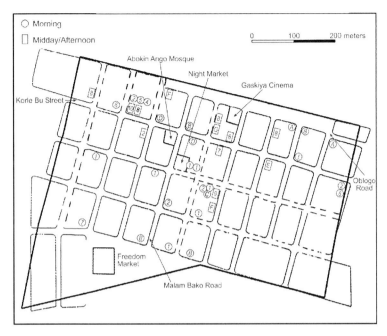

Figure 2. Foods Sold in Sabon Zongo, Morning and Midday/Afternoon

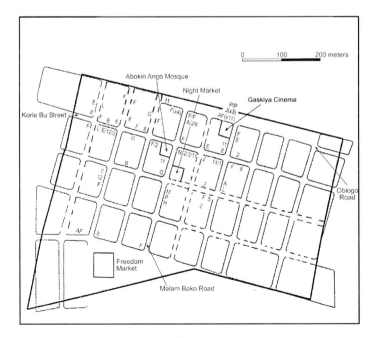

Figure 3. Foods Sold in Sabon Zongo, Late Afternoon/Evening

community. Far more food is sold at night than during the day. This is the time when men can hang out and socialize with friends, and how much more so over food. More exclusively Hausa chop is sold in the morning and midday. The kebabs and grilled side of lamb, the various kinds of bean cakes, the *fura* and soups–these are sold at the times that one would eat them in northern Nigeria. The kebabs, eaten all over Ghana, attract non-Hausa as well, as most people believe that the *zongos* have the best kebabs. The Muslims are the ritual slaughterers and the meat they grill is recognizable–it is only lamb or beef, never pork, cat, or dog. All Muslims know that the grilled meats sold here are not taboo.

There are also "floaters" who carry food on their heads to hawk, and all of the prepared food they sell is Hausa–*pinkaso, k'osei, k'afa, zogala* (spinach pounded with peanut butter and fresh pepper)–or fresh fruit. Even they have their prescribed routes. At the Night Market, on the other hand, most of the food sold is southern Ghanaian–overwhelmingly *kenkey*.

As in many of Accra's communities or neighborhoods, Sabon Zongo's market, Freedom Market, is open daily. In 1995, there were ninety-nine stalls–not one of them occupied by a Sabon Zongo Hausa woman. Many Hausa women engage in what the Yoruba call *jojoma* (always, i.e., every day): that is, they give something to the market women to sell for them and then return daily to be paid as the items are sold. What is interesting is that the Hausa women traders do not sit in the market. In northern Nigeria, the vast majority of the Muslim Hausa observe female seclusion; in Sabon Zongo this has never been the norm. But while women may be out in public and traversing the public space, they operate in separate social networks from men, and women do not sit and sell in the public market. Yet many of the Hausa women do sell outside–"*a bak'in k'ofa*," at the opening of their house, "because of *kunya*" (modesty); they believe they are maintaining their modesty by staying out of the public eye, even as they participate in public commerce, often by the side of the public street. "You will never find a Hausa woman selling in the market–only in her doorway–because of *kunya*." (Interview with Rukaya Barko February 3,1995). The Sabon Zongo Hausa women who sell food in the street redefine this as being less in the public eye than if they were selling at Freedom Market. A number of women have been selling regularly on Korle Bu Street over the last decade or so. Of the fourteen I spoke with, all have social reasons for selling where they do–they sell near their own house or their *mother's* house if they live elsewhere in Accra, somehow out of the public eye. It is an interesting negotiation of space, enabling their participation in the public economic sphere (see Cooper 1997).

Moreover, like their sisters in northern Nigeria, they tend to sell prepared foods, or foods that do not need preparation to be eaten, such as oranges, while the traders in the market proper sell ingredients only. Ten of the fourteen sell prepared foods, eight of them specifically Hausa foods. But of interest to us here is that it is clear for all of them that they sell here (rather than elsewhere in Sabon

Zongo or elsewhere in Accra) because there is a market for what they sell. *Masa,* for example, is eaten on Hausa ritual occasions. The little children, who are socialized into the defining foodways, come to buy the cakes every Thursday evening, to eat on the Muslim Sabbath. One food trader remarks that she "cook[s] to the taste of the people" (Interview August 5, 1995) in the area. Others note that they sell there because of the presence of many other cooked food sellers; even more important, they say, is the combination of particular foods – for example, selling *pinkaso* because it is a normal accompaniment to *koko,* or *masa* because it is taken with *koko* and one of the locally sold soups.

While men and women alike patronize the prepared food sellers, those who sit around eating in public spaces–for example, on benches near the Night Market and behind Gaskiya Cinema–are primarily men. Women buy the foods and take them home – for themselves, for their children, even for their husbands and fathers. In the Hausa homeland, men hang out together near the mosques and in the entrances to the family compounds. In Sabon Zongo, they do so by the corner barber, by the mosques, in the middle of an unpaved street behind Gaskiya Cinema, and in the alleyways playing cards. Similarly, it is men who hang around the *cincinga* (kebab) sellers, eating and chatting. Thus, by tracking the prepared food selling, one is also able to track patterns of social relations and exchange.

Conclusion

In this essay, I have posited that social ties and their cultural content, as well as the spatialization of local practices, are intrinsic to the satisfaction of physical needs and quality of life concerns, elements fundamental to a community's sustainability. I have explored this position through the example of Sabon Zongo, an urban community created in the early twentieth century as a planned town site in Accra, Ghana. To illustrate how the interaction of social, economic, and environmental factors has enabled Sabon Zongo to endure for almost one hundred years, I have focused on prepared food specialties and their spatial arrangement. The preparation and sale of food is particularly salient to the community's identity and the practice of everyday life, to the definition and sustenance of Sabon Zongo and the reinforcement of peoples' attachment. It is also crucial for basic survival, for there are many who cannot afford to cook. Community residents have maintained culinary traditions of vernacular culture, while also adapting or inventing new traditions or practices, and these have helped sustain the community.

This essay has looked specifically at the creolization of food consumption, as it dovetails with globalization, as one such practice that has helped sustain the local community. We see creolization in the face of a local cuisine and interpretations of Western *and* Hausa cuisine. In a sense, with their extensive trade networks, the Hausa are an earlier form of globalized community. They have carried their customs with them, influencing many. Some Sabon Zongo people cling to Hausa, rather than local or Western food, because it may reflect northern Nigeria. It reminds them of home, even though many have never been there. Some prepared foods, like kebabs, are associated with Hausa by southern Ghanaians. Outside of many bars in Accra, men sell kebabs. But one does not know what kind of meat is being grilled–it may be dog that the northern Ghanaian Frafra or Dagare eat, or cat for the southern Ghanaian Ewe. Thus there are those who go to the *zongo* because, following Muslim habit, they are guaranteed lamb or beef.

We must not forget the imagining and reimagining of tradition. Ethnicity is born of acknowledged difference, and ethnic foods are associated with particular eating communities. Because of the influence of living in southern Ghana, children may lose touch with Hausa foods; being able to buy them in the neighborhood maintains a connection. The prepared foods sold on the street are sold at the time of day one would consume them in northern Nigeria. And as people sit alongside one another eating them, they are engaging in and expressing a certain sociability. But just as nationhood and ethnicity may be imagined, "associated cuisines may be imagined too. Once imagined, such cuisines provide added concreteness to the idea of national or ethnic identity" (Mintz & DuBois 2002: 109). Some foods sold as Hausa may be Hausa creations in the diaspora; for example, *k'afa* is Kotokoli, and *wagashi,* which I also assumed to be a Hausa food, is actually part of Nupe cuisine, from the Nigerian area of Bida. Some Hausa foods sold in Sabon Zongo, like *wake,* are also sold outside, having been incorporated into Ghanaian consumption patterns. For several decades, Accra restaurants have served for brunch on Sundays *tuwon zafi* (known as t-zed), an interpretation of the Hausa food *tuwo,* accompanied by beer, which no self-respecting Muslim would drink.

We may note a number of points: (1) this community was founded by Hausa and it has perpetuated Hausa custom; (2) hence, the importance of ethnic, in this case Hausa, foods; (3) this community is located within a larger urban whole and ethnic mosaic, and its residents interact with the greater whole; (4) this has led to the reinvention of tradition, in the case of foods, creating some as "Hausa" that may or may not have been; (5) in some cases, there is improvement upon the traditional Hausa food; and (6) local Ghanaian foods have also been incorporated into the diet.

Observers of Hausa society have for years noted their disproportionate impact on clothing, language, ritual, religion and learning, economic affairs, roles, and offices (Schildkrout 1978). Clearly the food has also carried salience, for themselves and others, in a mundane but significant manner. It has helped socialize children into food preferences and cultural patterns more generally, and it has drawn Hausa scattered through the city to the community and attracted non-Hausa as well.

Non-Hausa and southern Ghanaians who are strangers to Accra and have moved into this community also find themselves at home foodwise. Thus, as the ethnic profile of the community is somewhat altered, Sabon Zongo is sustained. The foodways in Sabon Zongo have contributed to the society at large, enhancing processes of acculturation and hybridization, while helping to organize the community socially and spatially, and to sustain the community and people's attachment to it.

Notes

1. I am ever grateful to Muhsin Barko, my friend and assistant, without whose help my research in Sabon Zongo would have been far less successful.
2. Following the *jihad,* which banned alcohol, kola became the substitute.
3. Also referred to as "stranger," they are bound by language, cultural orientation, and education (see Margaret Peil 1979). They may include other ethnic groups from Nigeria, Burkina Faso, Niger, Togo, and so on.

References

Adamu, Mahdi. 1978. *The Hausa Factor in West African History.* London: Oxford University Press.

Agnew, John A., and James S. Duncan. 1989. "Introduction." In *The Power of Place: Bringing Together Geographical and Sociological Imaginations,* ed. John Agnew and James S. Duncan. Boston: Unwin Hyman.

Boeh-Ocansey, Osei. 1997. *Ghana's Microenterprise and Informal Sector.* Accra: Anansesem Publications Limited.

Brand, R. R. 1972. "The Spatial Organization of Residential Areas in Accra, Ghana, with Particular Reference to Aspects of Modernization." *Economic Geography* 48: 284-98.

Cohen, Abner. 1965. "The Social Organization of Credit in a West African Cattle Market." *Africa* 35 (1): 8-20.

Cooper, Barbara M. 1997. "Gender, Movement, and History: Social and Spatial Transformations in Twentieth Century Maradi, Niger." *Environment and Planning D: Society and Space* 15: 195-221.

Counihan, Carole M. 1999. *The Anthropology of Food and Body: Gender, Meaning, and Power.* New York: Routledge.

Dakubu, M. E. K. 1997. *Korle Meets the Sea: A Sociolinguistic History of Accra.* New York: Oxford University Press.

Dretke, James P. 1968. "The Muslim Community in Accra: An Historical Survey." Master's thesis, University of Ghana, Legon.

Fieldhouse, Paul. 1986. *Food and Nutrition: Customs and Culture.* London: Croom Helm.

Grant, Richard. 2002. "Foreign Companies and Glocalizations: Evidence from Accra, Ghana." In *Globalization and the Margins,* ed. Richard Grant and John Rennie Short. New York: Palgrave Macmillan.

Hannerz, Ulf. 1987. "The World in Creolization." *Africa* 57 (4): 546-59.

___. 1992. *Cultural Complexity: Studies in the Social Organization of Meaning.* New York: Columbia University Press.

Hart, Keith. 1973. "Informal Income Opportunities and Urban Employment in Ghana" *Journal of Modern African Studies,* 11 (1): 61-89.

Larbi, W. O. 1996. "Spatial Planning and Urban Fragmentation in Accra." *Third World Planning Review* 18 (2): 193-215.

Leeds, Anthony. 1973. "Locality Power in Relation to Supralocal Power Institutions." In *Urban Anthropology: Cross-Cultural Studies of Urbanization,* ed. Aidan Southall. New York: Oxford University Press.

Lisowska, J. 1984. "The Demographic, Social and Professional Structure of the Population of Accra between 1960 and 1970." *Africana Bulletin* 32: 113-29.

Mabogunje, Akin L. 1990. "Urban Planning and the Post-Colonial State in Africa: A Research Overview." *African Studies Review* 33 (2): 121-203.

MacClancey, Jeremy. 1992. *Consuming Culture: Why You Eat What You Eat.* New York: Henry Holt and Co.

Miller, Daniel. 1998. "Why Some Things Matter." In *Material Cultures: Why Some Things Matter,* ed. Daniel Miller. Chicago: University of Chicago Press.

Mintz, Sidney W. 1985. *Sweetness and Power: The Place of Sugar in Modern History.* New York: Penguin.

Mintz, Sidney W., and Christine M. DuBois. 2002. "The Anthropology of Food and Eating" *Annual Review of Anthropology* 31: 99-119.

Mintz, Sidney W., and Sally Price, eds. 1985. *Caribbean Contours.* Baltimore: The Johns Hopkins University Press.

Peil, Margaret. 1979. "Host Reactions: Aliens in Ghana." In *Strangers in African Societies,* ed. William A. Shack and Elliot P. Skinner. Berkeley: University of California Press.

Pellow, Deborah. 1985. "Muslim Segmentation: Cohesion and Divisiveness in Accra." *Journal of Modern African Studies* 23 (3): 419-44.

___. 2002. *Landlords and Lodgers: Socio-Spatial Organization in an Accra Community.* Westport, CT: Praeger.

Robertson, Claire. 1983. "The Death of Makola and Other Tragedies." *Canadian Journal of African Studies* 17 (3): 469-95.

___. 1984. *Sharing the Same Bowl: A Socioeconomic History of Women and Class in Accra, Ghana.* Bloomington: Indiana University Press.

Schildkrout, Enid. 1978. *People of the Zongo: The Transformation of Ethnic Identities in Ghana.* Cambridge: Cambridge University Press.

Globalization, Local Practice, and Sustainability in the High Plains Region of the United States

Janet E. Benson

Introduction

While researchers have examined economic and environmental factors affecting agricultural sustainability in the High Plains region of the United States, they have paid relatively less attention to sociocultural factors intrinsically linked to agricultural decision-making and the future of communities. Anthropologists can contribute to public policy decisions through a holistic perspective that takes into account both local practice, particularly the social relations in which agricultural practices are embedded, and an understanding of broader contexts (see Okongwu & Mencher 2000). This is not to deny the reality of local response, whether it entails resisting larger forces or embracing them. However, a broader perspective is needed to evaluate change over time. Regional, national, and international processes all shape the future of local communities.

Discussions of sustainability and communities of place have also sometimes treated farming communities as if they were socially and culturally homogeneous. However, this essay will argue that rural communities are not synonymous with farm families (see Scott et al. 2000) and in fact may be characterized by increasing diversity in terms of occupation, type of agricultural enterprise, and ethnicity (Benson 2001). The idea of local practice begs the question of exactly who is participating in regional culture; to what extent, if any, a collective identity exists; and how state and national policies shape local practice, as well as being shaped by it. The effects of globalization profoundly affect rural communities in the United States, as elsewhere, ranging from shifts in traditional markets to fluctuating energy prices and international migration. Highly mobile capital attracts equally mobile, and ephemeral, human populations. What changes have taken place in Kansas's agriculture during the last few decades?

To what extent are recent farming practices, based on unrestrained use of groundwater, cheap energy, and an expendable migrant labor force, likely to be sustained in the future? What are the consequences for rural communities of evolving practices? This essay will look at the interaction of social, economic, and environmental factors in southwest Kansas, center of the beef industry in the United States, where rural populations are both declining and becoming more ethnically diverse due to rural industrialization.

The Setting

To understand current sustainability issues in the High Plains region of the United States requires discussion of crop and livestock production, the meat-packing industry, and international migration. This essay focuses on the south-ernmost extension of the High Plains, particularly the southwest quadrant of Kansas surrounding the regional center of Garden City (see Figure 1). This com-munity was the subject of intensive ethnographic study during 1988-90 by a research team including the author, who has conducted several subsequent proj-ects in the area (Stull et al. 1990; Benson 1994a, 1994b, 1999, 2001). A recent pilot study focused on employer attitudes toward migrant labor use in dairying, swine production, and horticulture (2003).

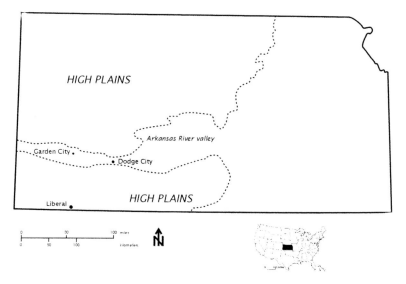

Figure 1. Southwestern Kansas Meatpacking Communities

The Agricultural Economy

Western Kansas is semi-arid, with rainfall averaging 14 to 22 inches per year (White & Kromm 1995: 277). As is the case elsewhere, water constitutes the critical, limiting factor for agriculture. Since the 1850s, the region's economy has been based on cattle ranching and farming. Settlement increased in the 1860s with the encouragement of the railroads and passage of the Homestead Act[1], although droughts led to periodic depopulation. Mechanization created other problems. Tractor-driven dryland farming, in addition to the Great Depression and drought, contributed to the "Dust Bowl" of the 1930s. The beginning of government aid (direct relief payments) also stems from this period (Gilson et al. 2001: 8-10). Contemporary agriculture in southwest Kansas incorporates local farming practices, developed since the 1960s, of cereal production (corn, wheat, sorghum, alfalfa) with center-pivot irrigation systems. These devices consist of an aluminum tube with attached sprinklers, supported by wheeled towers that move slowly in a circle. The towers can climb gentle slopes, making it possible to irrigate upland areas where gravity systems cannot operate (Hart 2003: 51). Intensive irrigation allows the production of water-thirsty crops such as corn and alfalfa, and produces much higher yields than rainfall-fed farming. In the 1970s and 1980s, for example, farmers could produce 115 bushels of corn per acre on irrigated land compared to 48 bushels with dryland farming methods (Gilson et al. 2001: 11). This is highly mechanized, large-scale agricultural production, supported by a modern agribusiness complex that includes seed suppliers, chemical producers, machinery dealers, and transportation networks (railroads and trucking). Contemporary agriculture in southwest Kansas is also closely integrated with rural manufacturing, specifically meatpacking.

Meatpacking

The development of irrigation-fed farming led to the establishment of cattle-raising and feeding operations that complement local beef-packing plants (slaughter and processing facilities). The restructuring of meatpacking since the 1960s and its move away from urban locations and toward rural communities near cattle and feed supplies has had a profound impact on rural society. Because cattle lose weight in transportation, it made more economic sense to move slaughterhouses near sources of feed supplies, rather than to continue shipping livestock long distances (see Broadway 1995; Skaggs 1986). The imposition of new environmental regulations and metropolitan growth have also forced beef packers out of the cities (e.g., Kansas City and Chicago) in which they were formerly located. Four new companies dominated beef packing during the 1970s, all of which eventually built or acquired plants in southwest Kansas: IBP, Inc.; ConAgra; Excel Corporation; and National Beef. ConAgra bought the Monfort plant in Garden City,

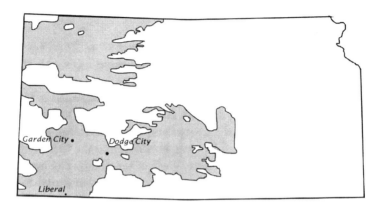

Figure 2. *Kansas Portion of the Ogallala Aquifer*

while in 1981 IBP opened the world's largest beef-packing plant in Holcomb west of Garden City (Stull et al. 1990: 2). IBP (now Tyson Fresh Meats, Inc.) is the world's largest supplier of beef and pork products, with sales of $16.9 billion in 2000 and more than 52,000 employees. In 2001 Tyson Foods, Inc., the world's largest producer of chicken, purchased IBP. The newly expanded company will control 28 percent of the beef, 25 percent of the chicken, and 18 percent of the pork market in the United States (Murphy 2001). Other plants operate at Dodge City and Liberal, southeast and south of Garden City respectively (see Figure 1). Today southwest Kansas is a major center for the U.S. livestock industry, with some of the largest feedlots and the world's largest beef-packing plant. Manufacturing, mainly meatpacking and value-added agricultural industry, brings $380 million per year into the local economy, while ranching, farming, and feedlots contribute $300 million and services $325 million, for a total of more than a billion dollars in revenue (Gilson et al. 2001: 14).

Water Matters

By early 2003, Kansas was in the third year of a severe regional drought. Farm income for the state as a whole had fallen from its five-year average net of $37,162 to $10,147 in 2002 (Manhattan Mercury 2003), spurring a continuing pattern of foreclosures, land consolidation and depopulation of many rural districts. Though this very serious drought affected even irrigated crops, local experts feel that periodic drought need not have long-lasting effects on agriculture. Groundwater depletion is a greater problem. Farm use is draining the Ogallala aquifer,[2] the major groundwater resource for Kansas and seven other states (see Figure 2). More than 3.6 million acre-feet are used each year in Kansas alone, but only 1.5 million acre-feet are recharged or absorbed by the entire aquifer. Depletion rates in southwest Kansas range from 10 to 70 percent, depending on aquifer depth and pumping rate, with water levels in some areas declining 100 percernt (Gilson et al. 2001: 1-2). More than half the land in this area, also known as Groundwater Management District #3 (GMD3) and including the counties in which Garden City, Dodge City, and Liberal are located, is now irrigated (White & Kromm 1995: 290). The Arkansas River, which has been used for irrigation in Colorado and Kansas since the late nineteenth century, is now largely dry. Western Kansas has little surface water today and depends heavily on groundwater resources (Sophocleous & Wilson 2000).

According to the Docking Institute report (Gilson et al. 2001), titled *The Value of Ogallala Aquifer Water in Southwest Kansas*, experts concur that the useful life expectancy of the aquifer is between fifteen and fifty years depending on local conditions. The Kansas Geological Survey defines "estimated usable lifetime" as the number of years until large-scale irrigation or industrial or municipal use becomes impractical, or the point at which water levels decline to a saturated thickness of thirty feet (Schloss & Buddemeier 2000). "Saturated thickness" refers to the "vertical thickness of an aquifer that is full of water," or the difference in feet between bedrock surface and the water table (Kansas Geological Survey 2000). This will vary locally depending on geological conditions and pumping rate. Some authors writing for the Geological Survey assume that sustainability is possible if average withdrawal rates can be reduced to somewhat less than the recharge rate (Schloss & Buddemeier 2000).

However, there is no agreement on possible solutions to the problem of sustainability. One response advocates moving away from current crops, diversifying, or even abandoning agriculture and reintroducing the American bison (the Buffalo Commons Thesis). A market approach, adopted by some local conservationists, assumes that low grain prices, declining water levels, and increasing energy costs will eventually lead to the end of irrigation. Others hope for technological solutions (Gilson et al. 2001: 15). For example, subsurface drip irrigation (SDI), pioneered by Kansas State University Research and Extension since 1989, uses half the water per acre of flood irrigation and 25 percent less than center-pivot sprinklers. However, the technology is more expensive to install than center-pivot systems, and farmers are unfamiliar with it (U.S. Water News Online 1999). To decrease the rate of aquifer decline, regardless of technology used, the number of acres must at least be held constant (as opposed to being increased). The introduction of center-pivot irrigation, which uses less water than flood irrigation, increased the number of irrigated acres and therefore depletion of the aquifer.

The Docking Institute report (Gilson et al. 2001), prepared for the Southwest Kansas Groundwater Management District, examined five irrigation scenarios in an attempt to predict costs and depletion of the Ogallala aquifer over a twenty-year period, assuming no major policy or economic changes. In Scenario 1, current practices would be followed (two-thirds of cultivated land under center-pivot irrigation methods, one-third using older flood irrigation methods that use 70 percent more water than center-pivot methods). In Scenario 2, center-pivot systems would replace the more wasteful flood irrigation method on all cultivated land. Scenario 3 would assume universal use of center-pivot systems but reduce water to all crops by 50 percent. In Scenario 4, water to all crops would be reduced so that yield becomes 90 percent of present yield. In Scenario 5, which the report describes as a "hybrid" solution, universal center-pivot use is assumed, but irrigation for corn is reduced by 50 percent, cutting yields by 10 percent. The result would be a near-zero depletion rate for the aquifer but at high monetary cost. The authors conclude that government subsidies will be required in the amount of $84,000 per section (640 acres) over a twenty-year period, or $4,200 per section annually in 1998 dollars, and the twelve southwest Kansas counties covered in their report include 2,618 irrigated sections (Gilson et al. 2001: 42-48).

This report clearly demonstrates that not only Scenario 5, but also all scenarios, including the one modeling current irrigation practices, require expensive government subsidies. The authors point out that if the value of subsidies is excluded, corn and wheat, together making up nearly 75 percent of the crops in southwest Kansas, have a negative cash flow of $150,000 per section in net present value. With subsidies included, income is about $170,000 per section, or $22,253,000 per year for irrigated land. Economists justify the subsidies by noting the multiplier effects of agricultural income in terms of money spent for goods and services (Gilson et al. 2001: 35-38). However, the fact remains that federal dollars are subsidizing the depletion of a valuable resource at a time of low crop prices and grain surplus.

Local experts on water management argue that subsidies have played an enormous role in the evolution of current farming practices. By the early 1980s, state records were often twenty years out of date, and on-site checks revealed that many wells were maintained only for the sake of government payments. As a result, the aquifer was rapidly depleted in areas such as Scott City, where the sustainable threshold has already been exceeded. By the 1980s, the only southwestern Kansas farmers not accepting subsidies were those about to retire. Unable to compete with other farmers operating under the subsidy system (handled by the U.S. Department of Agriculture and known as the "Farm Bill"), they were eventually forced out of farming. Because the government placed limits on individual holdings for subsidy purposes, farmers with fewer than two thousand acres felt they could not be self-sustaining. According to some informants, paper corporations and other subterfuges were used to obtain higher government payments for individual farmers. From the perspective of some water specialists, subsidies are the main cause of groundwater depletion. And because Kansas is dependent on them, this is a highly charged and politicized issue.

Given this background, there have been few serious attempts to reduce groundwater use until recently. However, in the 1970s, to counterbalance the state's emphasis on unlimited development, Kansas was divided into groundwater management districts, which theoretically have the ability to control water use. GMD3, for example, comprises twelve counties in southwest Kansas, including the three counties in which the major meatpacking communities of Garden City, Dodge City, and Liberal are located. Fifteen directors form its board, one from each county and three at-large, the latter representing municipal, industrial, and surface water use. Most of the members are farmers, many of whom have investments in local agribusiness. Board members are skeptical about the Docking report Scenario 5, which requires 50 percent less water use on corn. One of the problems is that tenant farmers are not free to make their own decisions about inputs, since retired owners living elsewhere do not understand the need to cut down on water consumption. Living on fixed incomes, these owners are unwilling to see yields decrease.

The GMD3 office places major emphasis today on enforcing water limits. Flow meters were placed on wells about ten years ago, and staff has a "zero tolerance" rule regarding water waste. Since the office has little money for policing and only five employees plus seasonal inspectors, court cases are the main form of deterrence; two hundred cases of noncompliance were identified in 2003 and all but one, which went to court, were resolved. Additionally, in 2004 board members dealt with the question of closing the district to new water appropriations, a highly contested subject. The board had already voted several times to close appropriations, but a public hearing held by the GMD3 proved a venue for vocal and well-publicized local opposition, including representatives from chambers of commerce and local legislators. As a result, the board met again after the hearing and reversed its previous decision.

Efforts are being made to develop a consensus regarding agricultural water use. The Kansas Water Congress is a very recent attempt to create a nongovernmental institution that will educate state legislators and the public about water issues. Incorporated in February 2002, the Congress will provide a forum for discussing water issues that may appear before the legislature and could serve as a conflict resolution forum. There are twenty-nine board members from twelve geographical regions in Kansas, representing not only agriculture but also various interests such as recreation, municipalities, and industry. One of the goals of the organizers is to promote broader, that is, not just local, views of current problems. Another function of the Congress will be to help the Division of Water Resources (the state water office) with planning. The Kansas Water Congress is modeled on the Colorado Water Congress, which has become a successful institution with good participation by "stakeholders" and legislators. The executive director of the Congress assumes that members will spend much of their time reacting to federal and state policies (such as the Clean Water Act) rather than in policy formation. However, to ensure as much consensus as possible, advocacy of particular policies must be expressed by a two-thirds majority vote. Given the strongly anti-government feelings among farmers, Congress organizers are apparently hoping to create a venue where conflicts over water use can be negotiated by water users themselves. The Congress is still in the developmental stage and its utility is yet to be demonstrated.

While marginal areas in the southern High Plains are switching to less water-intensive crops, water reserves are probably sufficient for thirty to forty years in some parts of southwest Kansas with current technology. Irrigated agriculture consumes by far the largest share of water; however, rural industries and towns with growing populations (e.g., Garden City, Dodge City, and Liberal) have increasing water needs. Contrary to assumptions that may be found in other societies, Kansas groundwater has been viewed until recently more as a commodity owned by individual farmers than as a public good managed by collective decision-making. This is due to state law, which varies throughout the United States but in the case of Kansas reflects a strong cultural emphasis on private initiative, economic self-sufficiency, and antagonism toward government regulation. Since virtually all of the state's groundwater rights have been appropriated, sales of water rights will become more common in the future. Water will also become more expensive as deeper wells are required (Lamm 2003), increasing the price of feed grain. State policy gives great autonomy at the local level to groundwater management districts, while state law emphasizes the right of prior appropriation (earlycomers have more water rights than latecomers). Irrigators therefore do not hold uniform interests, and a consensus on shared reductions has yet to emerge (White & Kromm 1996).

So far, this discussion has focused on crop and livestock production and their relationship to groundwater depletion. International migration introduces a new level of complexity in understanding local farming practices. Rural manufacturing, agriculture, and livestock production have become increasingly dependent on migrant and immigrant labor during the last two decades. Rural life has become more diverse in cultural terms, and the effects of globalization (such as the North American Free Trade Agreement [NAFTA] and foreign markets for meat and grain) are more keenly felt in the High Plains than previously.

Evolving Practices and International Migration

The use of migrant labor is not new to southwest Kansas. Mexican laborers worked on railroad maintenance and in farmers' fields during the twentieth century, particularly when Garden City had a flourishing sugar beet industry; the town's established Mexican-American population dates from this period (Stull et al. 1990: 50). Migration was minimal during the 1970s but by the 1980s, with the expansion of beef-packing plants in southwest Kansas, labor needs rapidly increased. Local unemployment was low, and plant managers knew that outside labor would have to be recruited. The work is dangerous, unpleasant, and relatively low paid, having been deskilled since the 1960s and moved to states with weak unions.[3] International linkages provide the needed labor force. Southeast Asian refugees and Central Americans, fleeing war in their own countries, provided many workers in the 1980s. This coincided with the opening of the world's largest beef-packing plant by IBP, Inc. in Holcomb, near Garden City, in 1981, and the renovation of other plants in Garden City and nearby Liberal and Dodge City (Stull et al. 1990: 2).

By the end of the 1980s and early 1990s, due to economic conditions in Mexico, deliberate recruiting by employers, and the passage of the Immigration Reform and Control Act (IRCA) in the United States, more and more Mexican migrants were arriving in southwest Kansas. Having legalized under IRCA after years of circular migration across the border, men brought their families to stay for the first time. Undocumented immigration also continued, contrary to the expectations of legislators but predictable considering job availability at the high-turnover plants and the strength of immigrant communication networks (Benson 1994b, 1999). At the same time, many Southeast Asian refugees left Kansas, having met their capital accumulation goals or being unwilling to continue under the plants' debilitating working conditions.

There is no satisfactory method of distinguishing recent Latino immigrants from established Mexican Americans since the federal census category of "Hispanic" lumps both together. However, Garden City's school district enrollment reports, one index of growth in the Latino population, demonstrate a steady (3-4 percent per year) increase in the proportion of Latino children since 1990. By September 2002 the proportion of minority students had reached over 65 percent (71 percent at the elementary level); comparable figures in September 2004 were 69 percent (74.6 percent at the elementary level). Nearly 60 percent of the district's students in 2004 were Latino or "Hispanic" (U.S.D. #457 2002, 2004). This upward trend shows no sign of faltering. Even the Christmas 2000 fire at Garden City's ConAgra plant, which closed the facility and put 2,300 people out of work, had only a temporary effect on the community's unemployment rate, which briefly rose from 4 to 12 percent. Swift Beef Inc. purchased the ConAgra plant in 2002 and plans to rebuild, which will attract more workers and their families (City of Garden City 2003; Fountain 2001). The large IBP plant at Holcomb currently has about three thousand employees; there are approximately 11,200 workers in packing plants in southwest Kansas. Although the workforce is diverse in terms of ethnicity, 75 percent of the employees in some individual plants are Latino (Christian 1998).

The last decade has seen a steady expansion of Mexican workers in other Kansas agribusinesses in addition to meatpacking. Recent federal programs that target large employers such as the packing plants have made it more difficult for undocumented Mexicans and Central Americans to work without papers of some kind (Benson 1999). This has tended to push newcomers into other jobs that are more difficult for the government to oversee. At the same time, labor demand for Mexican workers by employers has been increasing. The farm population is aging, high school and college students are no longer willing to engage in manual work, and married women are pursuing educational and career goals. Lower labor costs and the greater exploitability of migrant workers, particularly the undocumented, are probably also part of the explanation. Whatever the causes, non-Spanish-speaking farmers who never previously worked with Latino laborers are now employing them. With regard to western and southwestern Kansas, this trend is most evident in the case of the rapidly evolving dairy industry (Benson 2003).

As with other areas of agriculture, the Kansas dairy industry has undergone restructuring in recent years, and dairy operations are increasing in size. Before the mid-1990s, no dairies existed in southwest Kansas. By the end of 2000 there were eighteen large dairies in operation or planned for west and southwest Kansas, with 56,000 cows. The larger employers are most apt to employ migrant or immigrant labor, although even small dairies are also beginning to do so. The large dairies are being deliberately recruited at national dairy trade shows by a western Kansas economic development group, mostly representatives from local chambers of commerce, which touts the area's stable economy, central location, and "cost-effective labor force" (wKREDA 2006). The dairies and subsidiary industries increase water demand directly, and indirectly in terms of grain consumption, and the fact that they attract migrant labor forces adds to local population, with its attendant water needs.

Municipal Growth

Unlike many other small Kansas communities, the major towns of southwest Kansas (Garden City, Dodge City, and Liberal) have been growing, increasing municipal needs for water and other services. Garden City, for example, has mushroomed from a town of 18,256 in 1980 (Stull et al. 1990: 2) to a community of 30,000 today. As it was in the early 1980s, Garden City is still one of the most rapidly growing communities in Kansas. A regional center for southwest Kansas and the largest community in the western half of the state, the Kansas Department of Commerce has designated it "one of the areas with the highest probable percentage of growth over the next twenty years" (City of Garden City 2003). Population growth, like increases in irrigated acres, fuels increased water demand. In 2002, during the recent drought, the city's water use reached a record high of 78 percent of its appropriation (Ibid. 2003). A sizeable population also lives outside municipal boundaries, particularly immigrant farm workers, swelling Finney County's numbers to nearly 40,000 people. Ford County, the location of Dodge City, has more than 33,000 residents and Seward County, home to Liberal, more than 23,000. Of the remaining sixteen counties in southwest Kansas, none had an estimated population in 2006 of much more than 7500, while ten have lost population since 1971 (Policy Research Institute 2006). Large packing plants have located in Finney, Ford, and Seward, and much of the population growth stems from the lure of meatpacking and other agriculturally based employment. Growing numbers of "settled-out" migrants4 in Garden City and other southwestern communities create new challenges in terms of increased cultural diversity and greater requirements for health, education, and other services. Since many newcomers are undocumented, their lack of legal status creates new issues for themselves and rural communities (Benson 1999). At the same time, agricultural enterprises other than meatpacking are relying on migrant or immigrant labor, particularly Mexican workers. This trend seems unlikely to change anytime soon, although it adds another element of uncertainty to agricultural production; migration is heavily affected by U.S. laws and policies as well as economic and demographic trends in the countries of origin. Workers leave literally overnight, for example, if threatened by government raids. Meanwhile employers actively lobby for a continuous migrant flow at the state and federal levels.

Conclusion

In the following conclusion I discuss how this case fits into the field of common property and sustainability research. Arun Agrawal's (2003) review of work in this area provides a clear and insightful analysis that can be applied to southwest Kansas. Agrawal notes that the idea of the self-governing community as an alternative to market-oriented policies and the state is very attractive to policy-makers concerned with sustainability (2003: 243). Coercive conservation by states is expensive and communities can often manage resources more efficiently (2003: 246). However, in reviewing existing studies, Agrawal finds that authors have focused almost exclusively on the local, ignoring the external social, political, and physical environments. He points out that three important forces shape the contexts for common property institutions, all of which I have tried to address in this essay: demographic change (which affects the ability to create enforceable rules); the extent to which market forces have penetrated communities; and state policies. Local-state relations in particular require more careful exploration. "As the ultimate guarantor of property rights arrangements, the role of the state and overarching governance structures is central to the function of common property institutions" (2003: 250). Agrawal further suggests that it is necessary to look at power and micropolitics within communities, since power is basic to the process of resource management (2003: 257).

At this point I will return to the questions posed at the beginning of this essay. First, it is clear that rural places are no longer (if they ever were) culturally or socially homogenous. Within them, power differentials vary by class, generation, and ethnicity. Owners of large dairies, for example, are heavy consumers of water and may be willing to fight for their interests at the state level, while migrant laborers concentrate on daily problems of survival (Benson 1993). Among crop farmers, owners have different perspectives than tenants, while owners differ in the strength of their water rights due to the doctrine of prior appropriation. Residents of municipalities are likely to have different interests than farmers. Because groundwater levels can vary dramatically from one locality to another, and proponents of economic development must take an optimistic view of available water supplies, perceptions of the need for conservation vary widely also. Water use is likely to be an increasingly contentious issue in the future.

State and national policies also shape local practice. At the national level, the Farm Bill provides subsidies that allow "mining" of the aquifer, and government failure to promote energy alternatives to fossil fuel has led to rising and unpredictable gas prices. IRCA legalized several million Mexican workers and inadvertently encouraged new immigration to southwest Kansas. IRCA's employer sanctions are ineffective with regard to the larger employers, who continue to recruit labor directly or indirectly from Mexico.

In summary, current agricultural practices in southwest Kansas are based on lavish groundwater use, government subsidies, and a cheap, expendable migrant labor force. One implication of the cultural and social diversity discussed above is that developing a common understanding on groundwater and other issues will be difficult. Some observers argue that economic factors are moving far more rapidly than any institutions can evolve to influence farmers' behavior. The once cheap and plentiful natural gas reserve of southwest Kansas, which made mining of groundwater attractive, is being depleted more rapidly than the aquifer and may be available only a decade longer. Rather than buying directly from a gas wellhead on the farm itself, farmers are switching to diesel, and energy prices are rising monthly. Farmers are experimenting with cotton, a low-water input crop, while feedlots are shipping in corn as local supplies decrease. Moving away from feed grain production, however, will reduce the state's competitive advantages in livestock production.

If sustainability refers to long-term protection of basic resources while providing for human livelihood (cf. Redclift 1987), southwest Kansas clearly has serious sustainability issues. Reduction of groundwater use with present technology will require lower yields or a shift to crops other than those used for cattle feeding. More water-efficient technologies and shared reduction in water use are possible but require coordinated efforts that are difficult given current laws, water policy, and diversity among cultivators. Heavy dependence on migrant labor adds another element of uncertainty as well as increased demands on the aquifer due to population growth. Evolving local practices have long-term effects, not immediately evident to practitioners, of reducing local self-sufficiency and making farmers more vulnerable to international economic and social forces.

Acknowledgments

I would like to thank the anonymous interviewees who contributed to this paper, as well as David E. Kromm, Barbara Yablon Maida, and Carl A. Maida for reviewing an earlier draft. Any errors or omissions remain my responsibility. Donna C. Roper drew the maps, John L. Johnson provided assistance with census materials, and Orlen Grunewald answered questions on the beef industry.

Notes

1. The Homestead Act of 1862, which allowed settlers to claim 160 acres of free land, was enacted to stop speculative land purchases and to create a society of small property owners. In 1909, an expanded Homestead Act increased the acreage to 320 (Gates 1977: 111,124; Merrill 1996:434).
2. In western Kansas, the High Plains aquifer, a large concentration of sands, silts, gravels, and clays, is generally identical with the Ogallala formation, and the aquifer system is referred to as the Ogallala aquifer. In south-central Kansas (east of Dodge City) the aquifer is seen as geologically distinct and known as the High Plains aquifer (Buddemeier 2004).
3. That is, in states where union membership is not compulsory ("right-to-work" states). The unions that do exist often cooperate closely with management.
4. Those who have opted to remain more or less permanently in the United States rather than return to their home country, most commonly Mexico.

References

Agrawal, Arun. 2003. "Sustainable Governance of Common-Pool Resources: Context, Methods, and Politics." *Annual Review of Anthropology* 32: 243-62.

Benson, Janet E. 1994a. "The Effects of Packinghouse Work on Southeast Asian Refugee Families." In *Newcomers in the Workplace,* ed. Louise Lamphere, Alex Stepick, and Guillermo Grenier. Philadelphia: Temple University Press.

___. 1994b. "Staying Alive: Economic Strategies Among Immigrant Packing Plant Workers in Three Southwest Kansas Communities." *Kansas Quarterly* 25 (1): 107-20.

___. 1999. "Undocumented Immigrants and Meatpacking in the Midwest." In *Illegal Immigration in America,* ed. David W. Haines and Karen E. Rosenblum. Westport, CT: Greenwood Press.

___. 2001. "Vietnamese and Mexican Immigrants in Garden City, Kansas: The Changing Character of a Community." In *Manifest Destinies: Internationalizing Americans and Americanizing Immigrants,* ed. David W. Haines and Carol A. Mortland. Westport, CT: Praeger Publishers.

___. 2003. "Employer Perspectives, Integration Issues, and the Impact of Labor Migration on Rural Kansas." Paper presented at the American Anthropological Association Annual Meetings, San Francisco, California, November 15-19.

Broadway, Michael J. 1995. "From City to Countryside: Recent Changes in the Structure and Location of the Meat- and Fish-Processing Industries." In *Any Way You Cut It: Meat*

Processing and Small-Town America, ed. Donald D. Stull, Michael J. Broadway,, and David Griffith. Lawrence: University Press of Kansas.

Buddemeier, R. W. 2004. *An Atlas of the Kansas High Plains Aquifer.* Kansas Geological Survey. .

Christian, Shirley. 1998. "Hispanic Workers Revitalize a Town." *New York Times* January 29, A10.

City of Garden City. 2003. *City Manager's Annual Report.* April 15, 2003. .

Fountain, John W. 2001. "Needy Workers Wait for a Kansas Plant to Reopen." *New York Times,* July 10, 10.

Gates, Paul W. 1977. "Homesteading in the High Plains." *Agricultural History* 51 (1): 109-33.

Gilson, Preston, Joseph A. Aistrup, John Heinrichs, and Brett Zollinger. 2001. *The Value of Ogallala Aquifer Water in Southwest Kansas.* Prepared for Southwest Kansas Groundwater Management District. The Docking Institute of Public Affairs, Fort Hays State University.

Hart, John Fraser. 2003. *The Changing Scale of American Agriculture.* Charlottesville and London: University of Virginia Press.

Kansas Geological Survey. 2000. *An Atlas of the Kansas High Plains Aquifer: Glossary.* http://www.kgs.ku.edu/High Plains/atlas/glossary.mtm#s.

Lamm, Fred. 2003. Telephone interview with author, March 14.

Manhattan Mercury. 2003. "What's Drying Up in the Drought? Farming Opportunities.", *Manhattan (Kansas) Mercury,* February 16, A3.

Merrill, Karen R. 1996. "Whose Home on the Range?" *Western Historical Quarterly* 27 (4): 433-51.

Murphy, Patrick. 2001. "Meatpacking Brought New Way of Life." *Garden City Telegram,* 15 June.

Okongwu, Anne Francis, and Joan P. Mencher. 2000. "The Anthropology of Public Policy: Shifting Terrains." *Annual Review of Anthropology* 29:107-24.

Policy Research Institute. 2006. Kansas Data Archive, Policy Research Institute. Lawrence: The University of Kansas. Data from estimated population (Current Population Reports), U.S. Census Bureau.

Redclift, Michael. 1987. *Sustainable Development: Exploring the Contradictions.* London and New York: Routledge.

Schloss, J. A., and R. W. Buddemeier. 2000. *An Atlas of the Kansas High Plains Aquifer: Estimated Usable Lifetime.* Kansas Geological Survey. http://www.kgs.ukans.edu/HighPlains/atlas/atintr.htm.

Scott, Kathryn, Julie Park, and Chris Cocklin. 2000. "From 'Sustainable Rural Communities' to 'Social Sustainability': Giving Voice to Diversity in Mangakahia Valley, New Zealand."*Journal of Rural Studies* 16 (4): 33-44.

Skaggs, Jimmy M. 1986. *Prime Cut: Livestock Raising and Meatpacking in the United States, 1607-1983.* College Station: Texas A&M University Press.

Sophocleous, M. A., and B. B. Wilson. 2000. *Surface Water in Kansas and Its Interactions with Ground Water.* Kansas Geological Survey. http://www.kgs.ku.edu/HighPlains/atlas/atswan.htm.

Stull, Donald D., Janet E. Benson, Michael J. Broadway, Arthur L. Campa, Ken C. Erickson, and Mark A. Grey. 1990. *Changing Relations: Newcomers and Established Residents in Garden City, Kansas*. Report No. 172. University of Kansas: Institute for Public Policy and Business Research.

USD #457. 2002. Enrollment by School, Minority, Sex. September 20.

___. 2004. Enrollment by School, Minority, Sex. September 20.

U.S. Water News Online. 1999. "Interest in Sub-Surface Drip Irrigation Growing Among Western Kansas Farmers." October. http://www.uswaternews.com/archives/arconserv/9intin10.html,

Western Kansas Rural Economic Development Alliance (wKREDA). 2006.

White, Stephen E., and David E. Kromm. 1995. "Local Groundwater Management Effectiveness in the Colorado and Kansas Ogallalla Region." *Natural Resources Journal* 35 (92); 275-307.

___. 1996. "Appropriation and Water Rights Issues in the High Plains Ogallala Region." *Social Science Journal,* 33: 437-50.

Quality of Life, Sustainability, and Urbanization of the Oxnard Plain, California

Barbara Yablon Maida and Carl A. Maida

Sprawl, Suburbia, and the Politicized Landscape

The latest unplanned, random, urbanizing landscapes no longer adhere to traditional categories, such as urban, suburban, town, or rural. Agricultural land is being used for these landscapes, which are characterized by dispersed industry, homes, and stores. Daly and Cobb (1989) suggest that this use of agricultural land has its roots in the maximized profits of industrialization; more productive farms would require fewer farmers. Price does not take ecological or social consequences into account. This leads to an unsustainable model and the expedient of a land sell-off. Ultimately, in spite of the technology and intensification lavished on the production of food and fiber crops, the land itself is being degraded beyond repair; maintaining agricultural land is becoming an example of diminishing returns. Wendell Berry (1977) has written that the *culture* of agriculture is being lost and, once lost, will be irrecoverable.

Not unlike earlier planned suburbias, this form of sprawl promises the right of access to "pristine countryside," albeit with a simultaneous disregard for the countryside's degradation (Marx 1991). Sprawl contributes significantly to the reality of fewer farmlands; in many counties nationwide, even the excellent quality of the soil is not a deterrent to development (Sorensen and Esseks 1997). Development patterns invariably require more infrastructure, and so infrastructure is created and capitalized by an economic nexus of "outside" interests from the surrounding metropolis and beyond. Hence, the terrain has become politicized within California's agricultural regions. Supporters and detractors of growth each defend their claims and their often contradictory notions of community, land use, and sustainable development.

The Regional Setting

Ventura County has a population of some 820,000 and sits on the edge of the Los Angeles metropolitan area. The county's urban areas are, for the most part, separated by agricultural or open space lands, so they have not grown together into the amorphous sprawl of development that is common in the Los Angeles basin. Although Ventura County was not the direct target of as much growth pressure as its neighboring counties, it nevertheless took several steps to insulate itself through voter-approved limitations on land use. Because conventional land use controls were viewed as inadequate to the task of either slowing growth or preserving agricultural land, voters passed the SOAR (Save Open Space and Agricultural Resources) initiative in 1998, which limits the expansion of cities and urban development of remaining unincorporated areas until 2020 (Fulton et al. 2001). Between the county, ten discrete cities, and citizen-sponsored initiatives, a number of unique measures have been enacted to avoid the area becoming an extension of Los Angeles. These measures seek to maintain the rural/farming character of unincorporated areas while allowing each of the cities to evolve into mature urban centers.

SOAR in Ventura County had antecedents in Napa County (1992) and the city of Ventura (1995). Between November 1998 and January 1999, many cities of Ventura County voted on versions of the initiative. Seven out of ten cities immediately passed the City Urban Restriction Boundaries (CURBs) measure. The Ventura County version was considered the strictest set of growth controls ever voted upon in southern California. The SOAR initiative highlighted the role Ventura County plays within the region and the challenges created by the county's land use policies and government organization. However, residential, commercial, and industrial encroachment on the county's agricultural land appears to be proceeding in spite of SOAR. This may be due to such behavioral factors as residents' unwillingness to accept high-density housing, carpooling, and reduced water usage. Though presented as an issue largely about agricultural land, there remains an ongoing paradox, namely that of opposition to SOAR by many farmers and agricultural groups. The apparently limited success of SOAR may also be attributed in part to the continuing fragmented quality of individual environmental impact studies, which do not draw cumulative impact conclusions for an entire region. There is, however, much more to this. When studying parcels of land slated for development since November 1998, one sees that considerable acreage has been scheduled for development, from the years prior to SOAR.

Oxnard is a site of particular interest for an understanding of urban development pressures in Ventura County, as county agricultural production is centered here. There is the profound and poignant conflict between commercial, residential, and agricultural land uses in the city. Oxnard's size exceeds that of other county cities, as does its large population disparities between class and ethnicity. Finally, the fact that Oxnard is a coastal city and is thus considered prime real estate puts the city under intense development pressures.

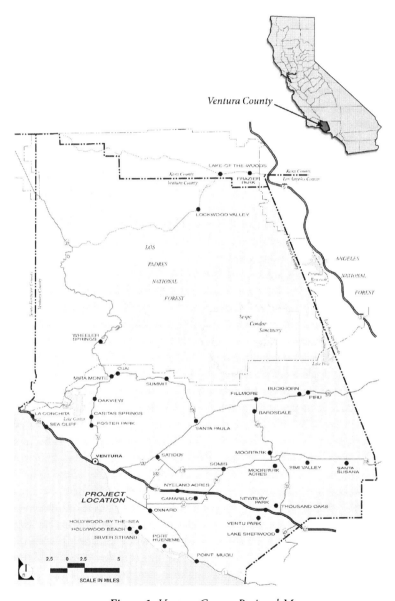

Figure 1. Ventura County Regional Map

The city of Oxnard is located in southern California, on the Oxnard Plain, in the southern half of the county of Ventura, which is halfway between Los Angeles County and Santa Barbara County (Figure 1). First developed commercially in the 1860s as an agricultural area (the Oxnard brothers grew sugar beets), the Oxnard Plain has also been the site of oil prospecting. Soils are extremely fertile, with natural drainage, and the climate is of a type characterized as Mediterranean. Population has increased at the same pace as agricultural production, which, in turn, depended upon ease of access to transportation. Incorporated in 1903, Oxnard's population has grown to over 125,000 people within approximately twenty-four square miles. Residents within the unincorporated areas of Oxnard bring the total population to approximately 180,000.

Oxnard can be reached by highway, railroad, air, and water. The deep-water port is the only commercial seaport between the Los Angeles/Long Beach and San Francisco harbors. In recent decades, Asia has been a high-volume importer of area citrus. County growers can therefore ship their produce internationally from within Ventura County; the lower shipping costs provide an economic incentive for growers to stay within the county, as does the employment of longshoremen, clerks, and foremen.

The Oxnard Plain is bordered by mountains and the Pacific Ocean. The center of the city is an industrial core that has, over time, encroached on the agricultural land. Currently, agricultural land is protected by the 1998 ballot measure. Ventura County contains less than 5 percent of the five-county metropolitan region's population but contains a large portion of the region's remaining agricultural land and open space (County of Ventura 2000). The ultimate use of this land is of tremendous importance, not only to the county but also to the region as a whole. The increasing connection of the county to the metropolitan economy, as well as its existing open land, are of importance to both Ventura County and metropolitan Los Angeles, giving the county new strategic significance for the region, particularly in light of the region's continued growth and rapidly increasing population and economy.

The County's Historical Relationship to Agriculture

Ventura County has recorded regional agricultural acreage since the 1920s; though maximum farmed acreage peaked in 1944, subsequent land conversions and slow but steady decrease in acres under production have not prevented overall crop values from burgeoning during each of the last six decades. The county has some of the most productive farmland in the nation; this is reflected in the figures for agricultural sales (Alvarez 2002; Ventura Council of Governments 2000; Wack 1974). If one traces the total crop values for Ventura County during the time period of the greatest land development and population increase (1980-2000), one sees little change in terms of constant dollars (County of Ventura

2000: 3). However, it must be emphasized that in order to reach these figures ($484 million in 1980, $853 million in 1990, $1,047 million in 2000), it is necessary to use all available acreage, plant multiple annual crops on the same acreage, and price the crops competitively. In other words, there are fewer acres overall grossing approximately the same amount of agricultural revenue.

There is more than one view as to whether acres under production have been diminished. The California Department of Conservation has documented steady farmland acreage loss for the years 1992-2006 in Ventura County. The most

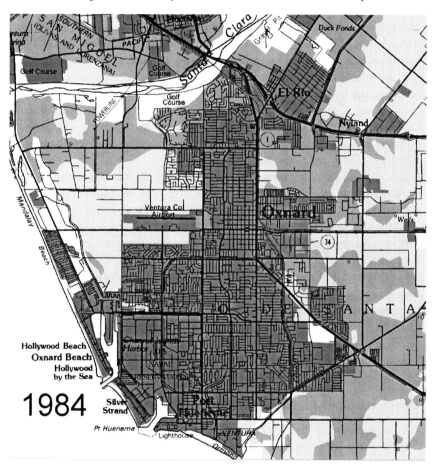

Figure 2. Existing Farmland Acreage and Soil Quality, Oxnard, California, 1984.
(Dark grey areas indicate residential and occupational development; medium grey indicates prime farmland; light grey indicates unique farmland or land used for orchards; and white areas indicate riparian, non-agricultural, or vacant land.)

Source: Farmland Mapping and Monitoring Program, California Department of Conservation.

dramatic losses occurred in 1994-96, with a net change of 1,433 acres (California Department of Conservation 2002; Ventura Council of Governments 2000). Time series maps constructed by the Farmland Mapping and Monitoring Program support this finding, as well as noting a shift in the proportion of prime acreage to soils of different quality and to developed land (Figures 2 and 3). Between 1994 and 1996, almost two-thirds of the county's lost agricultural acreage was classified as "Prime." In spite of this loss, more efficient use of lower-quality soils has been allowing for an increase in actual acreage under production.

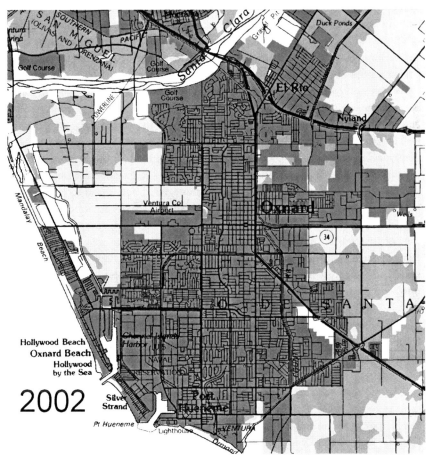

Figure 3. *Existing Farmland Acreage and Soil Quality, Oxnard, California, 2002.*
(Dark grey areas indicate residential and occupational development; medium grey indicates prime farmland; light grey indicates unique farmland or land used for orchards; and white areas indicate riparian, non-agricultural, or vacant land.)
Source: Farmland Mapping and Monitoring Program, California Department of Conservation.

This acreage includes soils classified as "farmland of statewide importance," "unique farmland," and "farmland of local importance" (Farmland Mapping and Monitoring Program 2006).

The estimated gross value of Ventura County agriculture for 2000 was in excess of one billion dollars, one-third from lemons and strawberries (Alvarez 2001; County of Ventura 2000). This increase in production is often accomplished by triple-cropping the acreage and is not without its own pitfalls. An oversupply of produce works to devalue county strawberries, which declined $34 million in revenue between 1999 and 2000. Growers on one-hundred-acre farms feel this loss to a much greater extent than do the larger growers because of their greater overhead costs; it is harder to price competitively unless a particular market niche is found. The landowners that ultimately sell to larger growers or to developers see the market as a losing one (Alvarez 2002).

What exactly is agricultural land? It should be relatively simple to chart the number of farm acres as a percentage of total county land. Likewise, field crops (grains, berries, vegetables, and fibers), pasture, and orchards (fruits and nuts) are generally well accounted for. Though much data is available on a yearly basis, precise documentation does not readily exist for all the land being used for the purpose of growing flowers and organic produce, for tree farms, nursery greenhouses, and apiaries, and for raising the beneficial insects used for integrated pest management.

Table 1 provides a cumulative view of land use conversions between 1992 and 2000, in the years leading up to and immediately following implementation of the SOAR initiatives. Though annual tables show net changes of zero acres in the inventoried area, the yearly exchange of land categories accounts for thousands of acres that move between different agricultural grades, "vacant or non-agricultural land," and "built-up land." Between 1984 and 1994, more than 11,000 acres of "important agricultural land" (7,500) and grazing land (3,600) were converted to development uses. Total agricultural acreage dropped from 345,000 to 334,000 acres in the county, and urban acreage increased from 77,000 to 92,000 acres (Ventura County Agricultural Land Trust and Conservancy 1996).

Regardless of whether the net change numbers add to zero, there is a steady drop in the number of agricultural acres. Conversions between agricultural land categories can go on only as long as extra land is available for exchange (Ventura County Agricultural Land Trust and Conservancy 1996). It is important to consider that this acreage may not shift categories indefinitely; using degraded land for agricultural or open space uses will not mitigate the inevitable trend toward urban usage. Available numbers can add ambiguity to the data; conclusions can be drawn that support both sides of the growth/no growth issue. By increasing the acreage of higher-value crops, there is not only a shift away from lower-value crops (such as sugar beets) but also a shift from level lands with adequate drainage to hillside acreage; this is a method that requires different irrigation styles.

Table 1. Cumulative Ventura County Land Use Conversions, in Acres, 1992-2000

Land Use Category	1992	2000	Net change
Prime farmland acres[1]	53,300	51,624	- 1,676
Important farmland[2]	125,298	122,942	- 2,356
Prime farmland as percentage of total important farmland	42.54	42.00	-0. 54
Grazing land	209,830	206,693	-3,137
Agricultural land[3]	334,928	329,635	-5,293
Urban and built-up acres	89,118	97,235	8,117
Other land[4]	129,457	125,144	-4,313
Water area[5]	3,016	3,939	0
Total inventoried area [6]	556,519	555,953	0[7]
Prime farmland as percentage of total inventoried area	9.6	9.3	-0.3[8]

1. In 1969, the acres considered to be prime agricultural were approximately 149,000.
2. Total Important farmland considers the acreage of prime farmland, farmland of statewide importance, unique farmland, and farmland of local importance.
3. Total Agricultural land is the sum of important farmland acreage, and grazing land acreage.
4. Other land is the designation assigned to riparian, nonagricultural/vacant land.
5. Slight year-to-year differences are the result of refining lake boundaries (California Department of Conservation 2002).
6. Total inventoried area is the sum of important farmland acreage, urban and built-up land, other land, and water area. Slight year-to-year differences are the result of adopting different areal projection methods, and/or refining lake boundaries (California Department of Conservation 2002).
7. Though the adjusted net change is zero acres, a total of 30,532 acres changed designation during this time period.
8. Using the 1992 figure for total inventoried area (556,519 acres), the percentage of Ventura County's prime farmland is calculated to have been 26.7 percent in 1969 (148,148 acres); thirty years later, the percentage has dropped to 9.3 percent (51,624 acres).

Though it is much more common for farmland to be converted (rezoned) to urban or built-up land, there are some cases where marginal land is upgraded to prime or unique land; in one extreme case for 2000, urban or built-up land was converted to prime farmland. North of the city of Oxnard, adjacent to a gravel pit, there are now two hundred acres in a prime land agricultural category (Farmland Mapping and Monitoring Program 2006). However, zoning controls do not alter the inevitable scenario of market values set by assessors: the "highest and best use," a term used by land use planners, has little to do with farming and affordable property taxes and much to do with development.

The land profile of the county, as expressed by the data in Table 1, reveals several patterns. The yearly exchange of land categories accounts for thousands of acres, moving between different agricultural grades, vacant or nonagricultural land, and built-up land. In the period between 1994 and 2000, year-to-year losses of agricultural land/other land precisely match the number of acres gained by urban/built-up land. The net change still records zero acres—none have slipped through the land survey—but the total of converted acres exceeds over thirty thousand during just six years. That amounts to almost eight square miles each year.

Environmental Perception as It Translates to Quality of Life

What about the perception of the residents, compared with the reality of development? Attitudinal data were gathered by a survey administered to a sample of 116 county residents (Maida 2002). Many who voted for the slow-growth measure considered their personal involvement and responsibility to be discharged on the morning after the ballots were counted; pro-development residents have stayed more consistently informed and involved with the land use issue. The structure of an attitudinal survey sought to downplay—as much as possible—the conceptual entity of SOAR as an influence on quality of life, emphasizing instead the perceived residential satisfaction of people living in Ventura County. If SOAR were to be cited as causal in any positive or negative way, it would be up to the individual to associate the initiative with open space, agricultural issues, and personal quality of life.

Aerial photographs formed the basis of comparison between actual land development projects and the perceived infractions of SOAR guidelines. Survey results and resident participation at meetings provided a subjective description of land use. Aerial photos were archived from the time that SOAR went on the ballot. Local aerial photographs of vacant land and maps of pre- and post-SOAR changes to Oxnard farmland provided data that formed the basis of a more objective land use description. These cartographic documents were studied concomitantly with the results of the attitudinal survey of county residents.

In the years since the SOAR initiatives appeared on local ballots, more than half of those surveyed stated that their quality of life has remained the same, more than one-third stated that it had improved in the past few years, and more than 10 percent felt that quality of life had worsened. When asked to speculate what the quality of life in their community would be twenty years from now, only one quarter believed that things would remain about the same; approximately one-quarter stated that it would be better or much better, while more than half viewed quality of life as being worse or much worse.

Subcounty comparisons were made to determine whether there were significant differences between the residential perceptions of quality of life, based upon individual town of residence. Eastern Ventura County is composed of towns having more day-to-day contact with the greater Los Angeles area, through both commercial and social networks. By contrast, the western part of the county is associated with the farming and coastal communities and is seen as having a greater remoteness from the metropolitan region. Nevertheless, very little of the populated southern part of the county may be described as rural; urbanization is proceeding throughout the area. When responses from those living in cities and towns in the eastern and western areas of the county were compared, eastern cities and towns showed significantly higher ratings for currently perceived quality of life and greater acceptance of agricultural land conversion. Western cities and towns were more negative, neutral, or conditional about these issues. However, there were no east/west differences in assessing the quality of life over the next twenty years or with regard to self-perceived satisfaction based upon residential town density.

The western Ventura County residents sampled had cited what they perceived as an accelerated number of development projects on and around the Oxnard Plain. To confirm whether this was so, observations were made as to measurable differences in residential development before and after passage of a local SOAR initiative. Comparisons were made among aerial photos that showed vacant and developed land, maps of Oxnard's vacant land parcels and zoning codes, and a list of development projects. According to this analysis, those projects that had reached early development at or before the time of the initiative's passage were making due progress but did not fall under SOAR guidelines. The residents did not have a clear understanding of the development timelines in their area, believing that the passage of SOAR would halt all future development, including projects already under way.

Much of the survey sample response was related to visible land conversion taking place locally. Oxnard residents in particular voiced their frustration with the early phase of SOAR initiatives. Many residents who equated visible agricultural acreage with quality of life gave other survey responses that revealed them to be largely unfamiliar with the operating principles of the SOAR initiative. Like those in other Ventura County cities, the Oxnard CURB delineates the outward point to which the city may develop; CURB lines are the reality behind the SOAR initiatives. In many instances, the same residents who had voted to curtail sprawl expressed apprehension that residential density, and more concentrated commercial districts, were the alternatives.

Other evidence of such misreading surfaced in that many of the respondents considered themselves to be living in a rural environment, when they in fact are dwelling in suburbia. Although the official designations consider the cities of Ventura County to be urbanized or urbanizing, it appeared that respondents held a false sense of the pastoral. There is a persistent belief that "continued urbanization will ruin local quality of life," especially in the towns of the western part of the county. Fifty years ago, Arthur Vidich and Joseph Bensman (1968: 79-81) made a study of how town residents respond to living among larger "agencies and institutions." The ambivalence that results from living in larger communities—the longing for a small-town romanticism, mixed with desire for access to modernity—may account for this view, shared by many respondents across the sample.

Nostalgic symbolism and pastoral idealism stand in sharp contrast to the more realistic view that the land in a market society is governed by contracts. The dominant economic ideologies of the United States, and other Western democracies, are built on market economies; land and nature are seen in economic or productive terms. This split between industrial and pastoral (or the mechanical and the organic) has continued to the present day. How can the old ideology and the "new American terrain" be reconciled? Marx (1991:75) considers the suburb to be a type of compromise, though he has always regarded suburbia to be "a modern, if somewhat debased effort to realize the pastoral ideal." Transportation has made this compromise possible, along with patterns of class and racial segregation.

The more uninformed residents are speaking in defense of an imagined landscape. Underlying these attitudes, based as they may be upon vague environmental perceptions, is a very real apprehension that the quality of life many residents are currently experiencing will not prevail in the near future. The more informed residents of Ventura County are familiar with the density and population pressures in neighboring Los Angeles County and are clearly not in favor of emulating those patterns of growth.

A Community-Based Indicator Design Process

In light of voter-approved limitations on land use, a group of informed residents has spearheaded a civic engagement process to adopt measures of the quality of life on the Oxnard Plain. These "quality of life" indicators attempt to measure sensitive features of the city's economy, social well-being, and environmental health near the beginning of the SOAR initiative. The goal was to see how Oxnard will fare as a growing and dynamic city once it runs out of land for growth and development within its city boundaries and cannot easily expand its borders into surrounding land since that land is governed by the SOAR initiative (Dagodag et al. 2001).

Each indicator was designed and placed in a set of interrelated factors so that they will likely show "links" or interconnections. One can expect, for example, that if Oxnard continues to build vast tracts of single-family dwellings instead of mixed-use, "livable" communities, the residents there will show certain unfortunate signs of personal malaise, even sickness, as they disengage from each other and see no reason nor find opportunities or places for active engagement with each other in a real community. The intent is to measure very human factors as they reveal themselves in quantifiable figures in the areas of health, economics, and environment. One outcome is a set of indicators that over time will show, in the simplest and clearest way possible, if Oxnard is a better or a worse place to live.

The Sustainability Council of Ventura County spearheaded the indicator design project. The council is a nonprofit, regional organization that establishes partnerships with businesses, organizations, governments, and individuals, including educators and university scientists, to conduct projects and programs that promote sustainable outcomes. Much of the council's work to date has focused on the problems of land use and the education about "livable" communities as an alternative to sprawl. The council was formed in 1995 following the first of several nationally televised town hall meetings on sustainability emanating from Washington, D.C., and sponsored locally at Point Mugu, a naval base being redesigned along sustainable principles in the post-cold war era. A core group of attendees organized themselves to produce subsequent annual conferences. They soon realized that a more formal organization was needed to continue their efforts year round. This led to the group's incorporation in 1997 as a nonprofit organization.

The council has served as the hub of community-based activities to promote dialogue and reflection upon sustainability issues that have surfaced as a result of the SOAR initiative. The council sponsored a local conference to promote the accurate tracking of various indicators that will inform the public and decision-makers if the county is moving toward or away from a sustainable future. From the meeting came the suggestion that a sustainability indicator should focus on civic engagement and that the council should take the lead in its design. This indicator would include such things as voter participation, attendance at civic organizations, volunteer time, attendance at public events such as street fairs, use of parks, libraries, and public places, and the general use of the community as a resource for personal enrichment and community connection. A focus group also took place with the Ventura County League of Women Voters that provided participants with an introduction to the indicator design process and an opportunity to discuss selected indicators.

The community-based forum and focus groups influenced the council's decision to concentrate on measures of the quality of life, namely urban density, health, and civic engagement in the city of Oxnard in the context of SOAR. A set of indicators was compiled to measure sensitive features of the city's economy, social well-being, and environmental health near the beginning of the SOAR initiative timeline of twenty years. These indicators emerged from a lengthy process involving council members' attendance at meetings and conferences on sustainable development and "smart growth," a stakeholder survey about the SOAR initiative and its efficacy, focus groups, and other community forums. As part of the indicator development process, existing sustainable development indicators from around the nation were reviewed. Indicators were formulated consistent with operational criteria that included the following: (1) indicators must be simple and easily understood by all residents of a community; (2) indicators must be premised on readily accessible information that requires little in the way of transformation or processing; (3) indicators must be valid measures that can be replicated periodically so as to gauge the direction of movement; and (4) indicators must be accurate depictions of conditions in Ventura County and especially for Oxnard since it is the city experiencing the greatest conversion of agricultural land and open space.

Civic engagement appeared as a recurring, but important, theme in indicators available from across the nation and similarly emerged in discussions with stakeholders and within the working group. Public participation is clearly one of the most important of the variables that relates to social equity and environmental justice matters. These indicators were, by far, the most difficult to agree on by everyone connected with the project. A further degree of difficulty was linked to finding data that was conveniently accessible and at the same time a reliable depiction of community involvement with quality of life issues.

The first public presentation of the indicators took place at a community meeting in Oxnard on the effects of land control measures in three related areas: land use and density, public health, and community involvement. More than forty residents attended the event, including the city mayor, in order to understand the complex purpose of sustainability indicators and to appreciate how such measures of the intangible quality of life can be used effectively in various forums. The discussion produced many suggestions of new indicators to develop and some suggestions about formatting or conceptualizing the indicators. A second community event in Oxnard, this time targeted to professionals in urban economic development, land use, and transportation, afforded an opportunity to disseminate selected draft indicators and invite comments from the three hundred participants.

As a result of these events, the council members began to think about shorter versions of the program that could be presented to various groups and organizations, and to list forums of access that are available to people for civic engagement around an issue, like SOAR, at the regional level (special interest groups, open forums; regional boards of public agencies, public agency-sponsored forums on regional issues, and regional organizations) and the local level (local chapters of national organizations, community-based organizations; emergent groups). The group also noted the "audiences" and public venues for civic engagement, including the Internet, phone trees, community cable television, school-based forums, the local public radio station, newsletters/publications of local organizations, and print media. Another discussion centered upon an Internet-based dissemination/feedback strategy, including brief policy statements and quick response surveys that would provide rapid-response "feedback" on efforts to reach the wider community. This would include an Internet-based consensus panel, including brief policy statements and quick response surveys, and may also include selected members of neighborhood watch and neighborhood councils, advocates, activists, clergy, elected officials, and interested residents who have responded to draft versions of environmental impact reports mandated in public land use planning and decision-making.

Sustainability indicators have a variety of uses, but those chosen by informed residents and council members were designed with three main purposes in mind. The indicators can be used individually to objectively measure progress or trends of urban development issues. For example, air pollution (violations) is a good indicator of reductions in vehicle emissions related to transportation changes. Increasing public awareness about the interconnections among the urban development indicators and the many quality of life outcomes is another major use of the indicators. For example, the mixed-use development indicator is directly connected to indicators of density and ease of traffic flow. These in turn influence parks and open space, peak stream flow, endangered species, beach closures, and air pollution. Related impacts include fewer pedestrian deaths, as well as pedestrian-associated health benefits of reduced cardiovascular disease and reduced childhood obesity. Reduced obesity results in fewer diabetes deaths. The increased pedestrian activity also fosters social interaction leading to stronger community civic engagement.

There are many other examples of interconnections between indicators. The idea that development choices have many ramifications for health and quality of life is one that should engage active public debate and further participation in revising or adding to the suggested sustainability indicators. This revision process and the ongoing data collection are key civic engagement issues that ultimately determine the usefulness of the indicators and the baseline indicator data obtained thus far. A third use of sustainable indicators is an increased realization of the many ramifications of development choices.

Sustainability Indicators, Civic Engagement, and Public Policy

Efforts to date have yielded a set of indicators with deeper roots in community life, with the potential for promoting local autonomy and a sense of uniqueness, as each community develops its own set of indicators (Dagodag et al. 2001).

Community residents, armed with increased understanding and indicator trend data, can more effectively engage in debate and influence urban growth policy; quantitative indicator data are more useful than general statements expressing concern. Linking causes and outcomes should also assist in prioritizing urban growth issues in the struggle for community consensus. Urban development in Ventura County may proceed as it has in most parts of the country, with construction of more residential and industrial areas connected by more streets and freeways, and with less and less farmland, and recreational and open space. The SOAR initiative may promote an alternative future with more preserved open space, mixed-use development and in-fill construction, more public transportation, and more public pedestrian-friendly areas.

Civic engagement is crucial to those decisions, and integral to the measured density and health outcomes. The idea that development choices have many ramifications for health and quality of life is one that should engage active public debate and further participation in revising or adding to the suggested sustainability indicators. In Lamont Hempel's words:

> Perhaps the most critical constraint on the development and use of sustainability indicators involves the role of ordinary citizens in their selection and interpretation. Deliberative democracy is, in many eyes, both a means and an end of the sustainable community movement. If deliberative democracy is conducive to the process of sustainability, and vice versa, it is important that citizens participate in the selection of indicators that will be used to evaluate their community and region. Although such involvement will sometimes lead to the inclusion of indicators that so-called "experts" regard as unscientific, irrelevant, or unreliable, to exclude such grassroots involvement may reveal, as clearly as any indicator, a basic cause of unsustainability—lack of civic engagement. (Hempel 1998: 29)

Consider that huge social inequities in income and educational opportunities produce commensurate and debilitating effects on the economy and the environment. In light of this, civic engagement on behalf of environmental justice will be necessary to mitigate the costs associated with increasing pesticide use and the associated health risks. Low-income communities have few ways to resist the degradation of local soils and water. Polluters exploit the lack of awareness and legal redress for such toxic effects and operate some of the most dangerous and environmentally damaging businesses, which often pay the lowest wages. An impoverished community is by necessity a poor guardian of environmental health. Huge disparities in income, which are increasing in the United States each year, threaten the security of society itself, as well as the economic strength of families and the productivity of workers.

Private Space and the Public Realm

How much can be taken out of the land without diminishing it? History is full of behavior that is grossly wasteful by today's standards. Until relatively recently, the world's limited human population mitigated this waste. Discounting the wholesale bison slaughters, it was marginally possible to follow a system of "cowboy economics" throughout the nineteenth century, without appreciable harm to North America's resources by "reckless, exploitative, romantic, and violent behavior" (Hardin 1993: 57-58). The fresh land that humans have always required for further expansion is now a scarce commodity; it is not difficult to see why ecology and economy, in spite of the common root *oikos,* relate uneasily. We have reached the point where what cannot be quantified cannot have value; the market has lost the ability to trace product back to source. Agriculture is not merely an industry; the free market cannot preserve "topsoil, the ecosystem, the farm" (Berry 1987: 123-28). Having a land ethic involves the preservation of landscape through a choice of a way of life.

What is the relationship between the realities of global marketing for agricultural products versus the local costs incurred by growers, and how does one evaluate the growers' choices of response? How have trade policies created domestic agricultural economies that favor risk (Gottlieb 2001)? The California urban real estate market has added additional pressure to the questions of farm location and productivity.

Overall, improvement of international farming technology has allowed for an expanded consumer market that, once created, is unlikely to be reduced for the benefit of individual domestic profits. Unlike participants in agribusiness, the choices for smaller growers are limited by available capital, and risky in terms of investment. The four-tier farming food chain, from low-value annuals such as grains, to high-value perennials such as orchards and vineyards, creates higher risk and the possibility of a more competitive product.

Ventura County's agriculture is close to the top of this hierarchy, with all attendant risks and benefits. Because of so many costs, profit is never high (Alvarez 2002, Hinojosa-Ojeda, Robinson & Moulton 1991). Being an efficient producer is not enough to ensure profitability, and so all types of farms are now being sold; the land's one-time real estate value far outstrips the return on continuing as an agricultural base (Blank 2000). According to the Ventura County Agricultural Land Trust and Conservancy, the shift to higher yield, higher value (and higher risk) crops "may be just the precursor to eventual urbanization" (Ventura County Agricultural Land Trust and Conservancy 1996: 92). Even in areas with runaway land prices, agribusiness may not be interested in family farms, unless they can be purchased in contiguous tracts. California's development decisions have steadily taken the land out of production altogether.

Since 1990, zoning law has reinforced a byzantine network within California's agricultural regions, such as the Oxnard Plain and Central Valley. Supporters and detractors of growth each defend their sides, along with a variety of often contradictory cost/benefit models. Current economic research for land use controls is often no more than a study of zoning (Fischel 1990). Measures of economic and political costs and benefits on building restraints resist being molded into a single-valued constraint on all building activity. More germane is whether land use control confers both benefits and costs, whether it is related to economically rational political activity, and whether its efficiency may be measured.

As land shifts irrevocably from one use to another, people tend to accept alterations in their physical environment as routine. Part of the problem seems to be a short-term attachment to any particular area—people come and go too fast to remember things as ever being different. The geology of California and other physical elements conspire in this; earthquakes, floods, and fires shift landscapes until they are unfamiliar. In places such as California's Great Central Valley, though the human tie is still visible, the region is undergoing changing uses as a result of the twin pressures of urbanization and economic development (Arax 1999). We sacrifice acreage that we have never stood upon and hope that elsewhere a similar piece of land will be permitted to continue producing a crop that links us to our imagined past. In this past, we keep a remnant of meaning, even at highway speeds.

Sustainable Development: Past Actions and Future Consequences

The fields of the Oxnard Plain have become a bellwether for checking on the fitness of the area. For both residents and casual observers, there is a very real connection between physical evidence of an agricultural economy and overall well-being. The advent of residences on the plain and recent conversion of highly visible croplands were the reasons that the 1998 SOAR initiative passed so handily in Oxnard and elsewhere. The "greater economic return" from the area is unchallenged; what is questionable is the certainty of "increased productivity in terms of land efficiency." Which brings the discussion back to the enterprises for which the SOAR initiative was supposedly written. Farmers in Ventura County did not instigate the initiative; unlike their cohort in Napa County, members of the local agricultural interest groups view this ordinance as detrimental on several levels. Not least is the fact that SOAR prevents the sale of their land for nonagricultural purposes. The freedom to decide whether a farm parcel has run its course has been taken out of the hands of those who have the most at stake. Land efficiency is not only about dollars and cents, but it must ultimately come down to certain realities. Tax base, civic events, and population densities are measurable standards. Open space parks would not result in short-term "economic productivity … and an increased revenue base" (City of Oxnard 1985: 200) and would have to be planned so as to limit human health risks related to liminal areas at the wildland-urban interface and development threats to agriculture at the urban fringe.

This essay raises a number of issues, most having at their core the concern of population growth and public policy. Much of the debate about the Oxnard Plain overlooks the needs of farm workers, recent immigrants, and those living at or near the poverty level. Class boundaries as well as growth boundaries therefore require careful study. For well over a century, preservation along with the management of labor and land development was the domain of urban elites. Before the union movement galvanized the lay public toward labor issues in the late nineteenth century, land was the primary source of tension between elites and ordinary people. More than thirty years after the first Earth Day, in these postindustrial times, land has again become a contested terrain.

However, abstract notions, such as "ecotopia," are not practical solutions to this tension over land use. To mitigate the effects of population pressure on the Oxnard Plain, higher-density housing needs to be built somewhere, which will be disconcerting to those in proximity. Those critical of sprawl as a growth pattern will need to accept some form of density as a necessary alternative and realize that higher-density housing is the most common solution.

People across the socioeconomic spectrum would benefit from education on land management policies. Though SOAR passed in some cities with greater than 60 percent of the vote, it is clear from responses to open-ended survey questions that many held a distorted view of the elements and implications of the voter-driven initiative. Clearly, this is not an issue to be solved solely at the polls. However, in the current climate of privatization, where complete information that speaks to both sides of an issue is difficult to obtain, it is unclear how land use policy can be left to a minimally informed public or to technical and political elites.

This essay adds to the discourse on a question first voiced at the end of World War II: What is the fate of open space in the face of sprawl? The "progress" inherent in the postwar boom produced consequences that are still being felt, such as settlement patterns that do not take a cumulative or a long view (Rome 1998). Commenting on the destruction of open space to create the suburbs at the end of the 1950s, William H. Whyte, Jr., warned that "much more of this kind of progress and we shall have the paradox of prosperity lowering our real standard of living" (Whyte, 1993: 133). At the end of the 1990s, when SOAR was enacted, this same paradox has continued to frame the debate over urban growth boundaries. Sprawl is a development ideal and a consequence of population pressure; open space is an environmental ideal, one that ameliorates the effects of sprawl (Duany, Plater-Zyberk & Speck 2000). The tensions between these two ideals and the competing concerns of abstract, popular ecology and local government, planning agencies, and an informed public must address reasoned strategies for open space preservation.

When public expectations are translated into policy initiatives, such as SOAR, that will influence local planning decisions for a generation, greater efforts will clearly need to be made to inform the public. First, attitudinal factors underlying proposed local ordinances should be interpreted in view of objective markers of change, such as aerial photographs and vacant land surveys. Second, residents' civic enthusiasm and willingness to impose urban growth limits need to be tempered by a thorough understanding of the parameters of urban development, including zoning codes and land ordinances. Finally, a fuller understanding of quality of life issues, such as community attachment, residential satisfaction, and aesthetic values, will need to inform planning decisions and local initiatives on behalf of preservation of open space and agricultural land.

Acknowledgments

Thanks to Tim Dagodag, Todd Collart, and Robert Chianese for their guidance and direction during our fieldwork, to Denis Cosgrove and J. Nicholas Entrikin for their comments on the sections of this paper presented at the Cultural Geography Methods Workshop, UCLA Department of Geography, to Regan Maas for making the maps print-worthy, and to Ken Meter and Marlene Grossman, who provided close readings and comments on earlier drafts.

References

Alvarez, Fred. 2001. "Crops Again Top $1 Billion." *Los Angeles Times,* June 27, B1, B11.
___. 2002. "County Strawberry Harvest Is Ripe for a Record Season." *Los Angeles Times,* February 12, B1, B11.
Arax, Mark. 1999. "Putting the Brakes on Growth." *Los Angeles Times,* October 6, A1, A18-19.
Berry, Wendell. 1977. *The Unsettling of America: Culture and Agriculture.* San Francisco: Sierra Club Books.
___. 1987. *Home Economics.* San Francisco: North Point Press.
Blank, Steven C. 2000. "Is This California Agriculture's Last Century?" *California Agriculture* 544 (3): 23-25.
California Department of Conservation. 2002. www.consrv.ca.gov/DLRP/index.htm.
City of Oxnard. 1985. *Final EIR for Oxnard Town Center 1984-1985,* vol 1.
County of Ventura. 2000. *Annual Crop Report 2000.* Santa Paula, CA: Office of the Agricultural Commissioner.
Dagodag, W. Tim, John Schillinger, Carl A. Maida, and Robert Chianese. 2001. *Impacts of Urban Growth Limits on Environment and the Quality of Life.* Hayward, CA: California Urban Environmental Research and Education Center.
Daly, Herman E., and John B. Cobb, Jr. 1989. *For the Common Good: Redirecting the Economy Toward Community, the Environment, and a Sustainable Future.* Boston: Beacon Press.

Duany, Andres, Elizabeth Plater-Zyberk, and Jeff Speck. 2000. *Suburban Nation: The Rise of Sprawl and the Decline of the American Dream*. New York: North Point Press.

Farmland Mapping and Monitoring Program, California Department of Conservation. 2006. www.consrv.ca.gov/DLRP/fmmp/time_series_img/index.htm.

Fischel, W. A. 1990. "Four Maxims for Research on Land-Use Controls." *Land Economics* 66 (3): 229-36.

Fulton, William, Chris Williamson, Kathleen Mallory, and Jeff Jones. 2001. *Smart Growth in Action: Housing Capacity and Development in Ventura County*. Los Angeles: Reason Public Policy Institute.

Gottlieb, Robert. 2001. *Environmentalism Unbound*. Cambridge, MA: MIT Press.

Hardin, Garrett. 1993. *Living Within Limits: Ecology, Economics, and Population Taboos*. New York: Oxford University Press.

Hempel, Lamont. 1998. *Sustainable Communities: From Vision to Action*. Claremont, CA: Claremont Graduate University.

Hinojosa-Ojeda, Raul, Sherman Robinson, and Kirby S. Moulton. 1991. "How the FTA Will Affect California Agriculture." *California Agriculture* 45 (5): 7-10.

Maida, Barbara Yablon. 2002. *Early Phase Efficacy of Slow-Growth Initiatives: S.O.A.R. in Ventura County, California, 1998-2001*. Master's thesis, California State University Northridge.

Marx, Leo. 1991. "The American Ideology of Space." In *Denatured Visions: Landscape and Culture in the Twentieth Century*. Stuart Wrede, ed. New York: The Museum of Modern Art/Harry N. Abrams, Inc.

Rome, Adam W. 1998. "William Whyte, Open Space, and Environmental Activism." *Geographical Review* 88 (2): 259-74.

Sorensen, A. Ann, and J. Dixon Esseks. 1997. *Living on the Edge: The Costs and Risks of Scatter Development*. DeKalb, IL: American Farmland Trust and the Northern Illinois University Center for Agriculture and the Environment.

Ventura Council of Governments. 2000. *The State of the Subregion: Measuring (Ventura County) Progress into the 21st Century*. Ventura, CA: Ventura Council of Governments.

Ventura County Agricultural Land Trust and Conservancy and the California State Coastal Conservancy. 1996. *The Value of Agriculture to Ventura County: An Economic Analysis*. Sponsored by the University of California/The Hansen Trust.

Vidich, Arthur J., and Joseph Bensman. 1968. *Small Town in Mass Society: Class, Power and Religion in a Rural Community*. Princeton, NJ: Princeton University Press.

Wack, Paul W. 1974. *The Land Conservation Act in Ventura County*. Master's thesis, California State University Northridge.

Whyte, William H., ed. 1993. *The Exploding Metropolis* (reprint of the 1958 edition). Berkeley: University of California Press.

Glossary

Area of interest: Each of the fourteen designated sectors in Ventura County, which can only contain one incorporated city (or a growing urban area that would incorporate in the future). Other cities cannot form, and county development cannot encroach upon the areas of interest. A paradox is created by the hold on county development after Proposition 13 reduced property taxes: retail construction (and sales tax) is of greater importance, but no retail taxes can be generated for the undeveloped portions of the county (Dagodag et al. 2001: 9-10).

Farmland: Important farmland categories for farmed areas where there are modern soil surveys.

- PRIME FARMLAND (P)
Farmland with the best combination of physical and chemical features able to sustain long-term production of agricultural crops. This land has the soil quality, growing season, and moisture supply needed to produce sustained high yields. Land must have been used for production of irrigated crops at some time during the four years prior to the mapping date.

- FARMLAND OF STATEWIDE IMPORTANCE (S)
Farmland similar to prime farmland but with minor shortcomings, such as greater slopes or less ability to store soil moisture. Land must have been used for production of irrigated crops at some time during the four years prior to the mapping date.

- UNIQUE FARMLAND (U)
Farmland of lesser-quality soils used for the production of the state's leading agricultural crops. This land is usually irrigated but may include nonirrigated orchards or vineyards as found in some climatic zones in California. Land must have been cropped at some time during the four years prior to the mapping date.

- FARMLAND OF LOCAL IMPORTANCE (L)
Land of importance to the local agricultural economy as determined by each county's board of supervisors and a local advisory committee.

Interim farmland categories for farmed areas lacking modern soil survey information and for which there is expressed local concern on the status of farmland.

- IRRIGATED FARMLAND (I)
Cropped land with a developed irrigation water supply that is dependable and of adequate quality. Land must have been used for production of irrigated crops at some time during the four years prior to the mapping date.

- NONIRRIGATED FARMLAND (N)
Land on which agricultural commodities are produced on a continuing or cyclic basis utilizing stored soil moisture.

Categories common to all maps

• GRAZING LAND (G)
Land on which the existing vegetation is suited to the grazing of livestock. This category was developed in cooperation with the California Cattlemen's Association, University of California Cooperative Extension, and other groups interested in the extent of grazing activities. The minimum mapping unit for grazing land is forty acres.

• URBAN AND BUILT-UP LAND (D)
Land occupied by structures with a building density of at least 1 unit to 1.5 acres, or approximately 6 structures to a 10-acre parcel. This land is used for residential, industrial, commercial, construction, institutional, public administration, railroad and other transportation yards, cemeteries, airports, golf courses, sanitary landfills, sewage treatment, water control structures, and other developed purposes.

• OTHER LAND (X)
Land not included in any other mapping category. Common examples include low-density rural developments; brush, timber, wetland, and riparian areas not suitable for livestock grazing; vacant and nonagricultural land surrounded on all sides by urban development; confined livestock, poultry, or aquaculture facilities; strip mines, pits; and water bodies smaller than forty acres.

• WATER (W)
Perennial water bodies with an extent of at least forty acres.

Five-county metropolitan region: Los Angeles, Orange, San Bernardino, Riverside and Ventura counties.

Prime class I and II soils: Designating an ability to grow almost any crop, California's total acreage under classification is approximately 100 million acres. Approximately 2 percent (1,630,000 acres) is classified as class I soil; approximately 5 percent (5,000,000 acres) is class II soil. Other areas designated as prime are Napa Valley, Salinas Valley, Santa Clara Valley, and Monterey Bay (Wack 1974: 8-9).

SOAR Initiative, City of Ventura, California: Save Our Agricultural Resources, passed in 1995, was patterned after Napa County Measure J, which was proposed in 1990 and upheld five years later in the California Supreme Court. This was the precedent that allowed General Plans to be amended by initiative. Some say that the difference between the Napa County and Ventura initiatives is that in Napa County, the growers were solidly behind the measure. In Ventura County, it is feared that growers will not be given the choice as to whether they can sell their land and leave the farm industry.

SOAR Initiative, County of Ventura, California: The Save Open Space and Agricultural Resources initiative (also known as Measure B) initially passed in 1998; Camarillo, Oxnard, Simi Valley, and Thousand Oaks joined the city of Ventura in approving local measures on the first ballot. Newbury Park, Moorpark, Santa Paula, and Fillmore subsequently passed the measure.

Sphere of influence: Within each area of interest, the sphere of influence demarcates boundaries outside of which a city may not annex land. This "ultimate city boundary" is nevertheless subject to changes by the Local Agency Formation Commission (LAFCO), consisting of city and county agents and citizens. Cities may also de-annex land acquired prior to the LAFCO guidelines and receive other unincorporated land in exchange; Oxnard did this in order to make its boundaries more compact (Dagodag et al. 2001: 10).

Zoning categories for agricultural lands: The minimum mapping unit for all categories is ten acres unless specified. Smaller units of land are incorporated into the surrounding map classifications (Farmland Mapping and Monitoring Program, 2006).

Part Three

SOCIAL CAPITAL, CIVIC ENGAGEMENT AND GLOBALIZATION

Linked Indicators of Sustainability Build Bridges of Trust

Kenneth A. Meter

Citizens who pursue environmental initiatives in community contexts face an *ecology* of social and economic issues that is extremely complex and often difficult for the outsider to penetrate. This fact alone suggests the need for local residents to be intimately and powerfully involved in partnerships with professionals who address sustainability concerns. Further, there is strong reason for professionals who work in community settings to maintain a detachment regarding their own professional expertise. Professionals tend to be specialists, while residents tend to be generalists: they are the experts on local *systems*. Frequently, academic or professional specialties contribute to a discourse that is narrower than is required to address the complexity of local issues. Dealing with single issues in isolation from each other may play a powerful role in creating models for action, but such a strategy is seldom effective in addressing complex issues, especially in rapidly changing conditions.

Several Minneapolis neighborhoods fronting the Mississippi River have worked closely with both professional and academic experts to take solid, practical steps toward building a more integrative approach. Central to these initiatives is a conviction that environmental, social, and economic issues cannot be separated. As partners in the first U.S. effort to engage residents in creating sustainability indicators for their own locales, residents of Seward and Longfellow neighborhoods built new bridges of trust, linked issues that had been fragmented, and defined pioneering indicators. These are not only important to their locale but also useful as tools for other locales globally.

Subsequently, leaders in a low-income community of North Minneapolis extended this approach by setting up a ten-year initiative to reduce poverty and build wealth in their community. Extensive public meetings developed a local analysis that poverty is systemic and must be overturned through systemic change. To build a more sustainable community, residents decided to create an overarching 'community service organization' that will coordinate the work of more than 170 existing community initiatives and their 170 external partners. Systemic evaluation techniques are intended to keep this work focused on systemic social change.

Neighborhood Sustainability Indicators Project: Seward Neighborhood and Longfellow Community in Minneapolis (1998-1999)

The Neighborhood Sustainability Indicators Project (NSIP) in Minneapolis emerged out of six years of preparatory work by environmental activists, who sought to place sustainability more firmly on the city's agenda. Leaders began to meet in 1992, strategizing how to create a "Sustainable Minneapolis" campaign. Early efforts to bring a strong labor voice into this discussion failed. After stops and starts, a group of environmental organizations framed themselves as the Urban Ecology Coalition (UEC). Holding a conference, they discovered strong citizen interest in environmental issues. A survey of these constituents identified priority areas of concern. Subsequent analysis determined that neighborhoods would be more responsive to engaging in sustainability efforts than would the city itself. Adopting measures of sustainability at the neighborhood level, UEC reasoned, might ignite greater attention to the concept at the grassroots level, ultimately inspiring regional efforts toward sustainability.

By 1998, UEC raised a small grant from the Minnesota Office of Environmental Assistance; a private donor matched these funds. Crossroads Resource Center (CRC) was selected to implement the effort—which, it turned out, was apparently the very first in the United States to engage neighborhood residents in defining indicators of sustainability for their own community. Two neighborhoods were selected to participate in the project. The first, Seward Neighborhood Group (SNG), was one of the more advanced neighborhood organizations in the Twin Cities of Minneapolis and St. Paul. A functioning organization for forty years, it had a budget of $400,000 (this has subsequently been reduced by funding cuts). With a population of seven thousand including a solid core of environmentalists, Seward also appeared to be an ideal candidate for the effort. In strong transition, Seward faces property values that are exceptionally high. Many houses are sold days before they are formally placed on the market. New homeowners are far wealthier than historic residents. At the same time, a new community of East African immigrants of very limited means has settled into apartment complexes in Seward. A second organization, Longfellow Community

Council (LCC), also participated. LCC is actually a coalition that spans twenty thousand residents living in four neighborhoods of South Minneapolis (Cooper, Hiawatha, Howe, and Longfellow). LCC was unable to make solid progress under the initiative due to internal reasons. Thus, this review of the neighborhood sustainability indicators project focuses on SNG's work.

Learning from Previous Initiatives

After studying a number of previous indicator projects.[1] CRC concluded that a wealth of solid technical material was available through both the academic literature and the Internet. We were guided in our analysis by the pioneering work of Virginia Maclaren of the University of Toronto. Maclaren (1996) set out a solid framework for distinguishing sustainability indicators from other types of indicators in the following way: (1) performance evaluation indicators are less about looking forward and more about assessing past efforts of a community or organization; (2) quality of life indicators tend to ignore the linkages among the varied issues neighborhoods face, focusing more on single-dimensional counts (see Besleme & Mullin 1997). Building upon these insights using the work of Maureen Hart,[2] director of Sustainability Measures, Inc., and former consultant to the U.S. Environmental Protection Agency, as well as John Kretzmann and John McKnight (1995) from the Center for Urban Affairs at Northwestern University, we defined sustainability indicators as holding the following qualities: (1) asset-based: determined in a process that begins by analyzing existing assets and addresses deficiencies later; (2) engaging to residents and other diverse stakeholders: defined with strong involvement by a diverse cross-section of residents and other stakeholders, with the benefit of professional assistance as appropriate, in respectful, mutual, flexible, and open decision-making processes; (3) express local values: measures progress toward neighborhood values adopted by local residents; (4) integrating: illuminates linkages among multiple issues and helps define integrated responses; (5) forward-looking: focuses on long-term future change, not evaluation of the past; and (6) distributional: works toward equitable distribution of resources, opportunity, and wealth, not only for the current generation but also for future generations.

Examining prior initiatives, we learned that several communities or regions had developed impressive slates of indicators. Still, at that time—and this has changed dramatically in recent years—few of these projects had engaged people at the neighborhood level in strategic planning. In some cases, professional experts had defined indicators externally, carrying on no conversation with local residents concerning local goals and making no effort to tap local wisdom from the history of local sustainability efforts. In other cities, regional indicators had been selected by a core of thoughtful and well-informed civic leaders, representing a far-sighted vision for the ecological health of the region. Yet the indicators selected were at such a large scale that citizens, who often can intervene most effectively at a local level, had difficulty translating regional goals into local action. Often, in fact, local people were unaware of that regional vision, nor had they been involved in framing it. Other indicator projects failed to recognize the linkages among issues that residents' groups faced: for example, the fact that a given economic development effort might increase transportation and housing costs for local residents did not appear to be taken into account.

Rooted in Long-Term Neighborhood Action

Our initiative defined "sustainability" to mean that people of our era can live well without asking future generations to pick up the tab. This is consistent with the definition of the World Commission on Environment and Development (also known as the Brundtland Commission (WCED 1987). Typically, neighborhood planning in Minneapolis has at best focused on short-term goals of five years or less. Exploring long-term visions challenged neighborhood partners to venture into new terrain. Residents felt they had been limited by bureaucratic constraints that had required them to consider only short-term goals. They eagerly rose to the moment. As Seward residents imagined a fifty- or one-hundred-year future, it became obvious that one of the obstacles to long-term planning was the press of short-term activity. With thirty-nine years' experience as a resident-run neighborhood organization, SNG was highly effective at accomplishing its objectives and expert at involving local residents in framing strategic goals—so successful, in fact, that the local housing committee worked more than full time accomplishing housing goals, while the economic development committee focused intensely upon commercial revitalization, and still other committees tackled issues of public safety, transportation, youth involvement, and so forth. However, never did these diverse committees work together to reflect on the interactions of their implementation efforts: there was no common answer to questions such as, how do our economic development investments help or hinder our housing programs? How might our investment in transportation networks reduce our public safety costs?

Reflecting on Local Experiences

NSIP became the crucible in which such issues were addressed. Delegates from diverse interest areas in Seward compared experiences and asked each other what made their work effective in promoting long-term stability. The discussions resembled those of a graduate field study seminar, with high-quality reflections on how to effectively intervene in a complex system, how to understand linkages among diverse issues, and how to produce data that would be useful for residents in shaping their future plans. Residents said they greatly enjoyed having this rare chance to reflect together and build a more lasting vision for their community. "Even if we never use a single indicator the process has given us so much," concluded the Seward board chair.[3]

Blending Local Experience with Citywide Strategizing

Local action in Seward was also aided by citywide strategizing, in three citywide roundtable meetings. Up to sixty-five residents, civic leaders, and technical experts engaged in far-ranging roundtable discussions in 1998, 1999, and 2000 that shaped the research and action strategy.[4] These roundtables proved exceptionally effective at identifying the major issues that had to be addressed at the neighborhood level. The roundtables also highlighted the fact that neighborhoods could not simply draw a wall around themselves and look inward, they also had to ensure accountability to external communities, funders, and investors. In turn, neighborhoods could use their self-defined indicators as a powerful tool in convincing external partners to address neighborhood goals.

Three Types of Indicators Defined

To respond to this blend of internal and external needs, CRC defined three kinds of indicators: systemic (also called "data poetry"), core, and background.[5] As the diagram (Figure 1) shows, these fall along a continuum. Systemic indicators do best at expressing linkages among local issues and are primarily for internal use. Core indicators are highly linked and seem appropriate for careful cross-neighborhood comparisons. Background indicators are less adept at defining linkages and may best serve external partners. All three seem useful in advancing neighborhood sustainability.

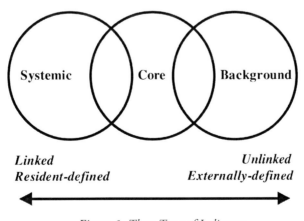

Figure 1. Three Types of Indicators

The overlapping circles in the diagram represent the fact that no hard boundaries exist between these three categories of indicators. Rather, they exist on a continuum. All three kinds of indicators quite naturally emerge in any indicator-definition process. Placing them on the continuum of linked vs. unlinked, and resident defined vs. externally defined, helped us understand the strengths and weaknesses of each.

Systemic indicators are those that capture the imagination of local residents, express in a concise way the linkages among several issues important to sustainability, and have the quality of transforming the discussion of the future of the neighborhood into a longer-term perspective. CRC believes these indicators are best defined by neighborhood residents who may solicit expert assistance in the process if they wish.

One excellent example of a systemic indicator is a Native American tribe in Maine that says one key core indicator is the balance of the wolf and moose populations. Since local hunters depend on moose as a meat source, local leaders reason that if the moose population is in balance with its predators, then they know the ecosystem is in balance, and indigenous people will survive.[6] Thus, these two species serve as indicator species. Sustainable Seattle's indicator of ecosystem health is the count of the salmon run in the Puget Sound watershed. A healthy salmon run, leaders argue, indicates that environment in the local watershed is relatively healthy. An optimal count means the water quality is high, erosion from farms and forests into streambeds is tolerable, and pollutant runoff is at safe levels. A high count also implies that local residents will have the opportunity to eat fresh fish, which can be a measure of nutrition. Further, the count suggests that a substantial fish harvest will help fuel the local economy.[7] Of course, opponents have argued that restoring a high salmon count by removing dams would interrupt shipping and reduce hydroelectric production, which may have negative consequences. Although it is a theoretical issue at the current time, it is also possible to have an overabundance of salmon, which would indicate imbalance.

During the Neighborhood Sustainability Indicators Roundtable in February 1998, there was a consensus that neighborhoods' most obvious role was to identify systemic indicators, based upon their own local goals. Each of the participating neighborhoods was invited to select five to ten such indicators. The systemic indicators defined by SNG include: (1) friendly space (places in the neighborhood that encourage residents to meet each other and build social connections with each other, measured with an original point system); (2) consumption by residents at independent local stores; (3) purchases from local vendors by local businesses; (4) number of residents who share skills or barter services with each other: (5) number of residents who volunteer for church or community service work; (6) number of residents who plan to stay in the neighborhood for a specified range of years; (7) number of bicycles traveling on key traffic routes compared to number of cars: (8) number of home-based businesses and resident-managed studio/office spaces; (9) percentage of residents earning a living wage: and (10) percentage of workers working inside and outside of the Seward neighborhood.

Core indicators—those in the middle of the three circles—reflect more detached research. Central to local sustainability, core indicators also suggest questions that could be addressed on a citywide or regional level. They are relatively highly linked, though in our experience not as closely as systemic indicators are. They fall at the intersection of wisdom held by external professionals and local residents. They may provide points of comparison across diverse neighborhoods, since they are more removed from local goals and capacities and may perhaps reflect a broader sustainability agenda.

Working closely with SNG, CRC defined twenty-five core indicators. These represent an effort to understand challenges that are broader in scope than those addressed by neighborhoods. Simultaneously, we attempted to root them in the experience of Seward residents but also connect them to regional concerns and make them useful for cross-neighborhood comparisons. One example of a core indicator was the affordability of housing. This turned out to raise an issue vital to the survival of SNG and central to regional policy discussion. At the time (an essentially new city council has been elected since this program was undertaken), city officials considered rising property (home sale) values to be a sign of success in civic life. To the residents of Seward, however, the official "indicator" that looked only at a linear rise in home sales was off the mark. These long-term residents recognized that after nearly four decades of working against great odds to improve the neighborhood, many community leaders could no longer afford to buy the homes they had worked so diligently to save. Their successes were literally being turned against them. A better measure of Seward's sustainability, they argued, was to compare the cost of homes to prevailing incomes in the neighborhoods and those nearby. By this measure, Seward was far less sustainable than the official view would acknowledge.

The third type of indicator proposed by CRC was a background indicator. One useful analogy here is to imagine actors in a play. If systemic indicators equate to the star actors of a play (key characters and themes) and core indicators constitute the other characters, then background indicators are the scenery. Not as central to the action involved in making the neighborhood more sustainable, they are nevertheless useful, especially for external use. Funders who want to compare conditions or results across neighborhoods, researchers who want to know when a strategy may work best in a certain kind of neighborhood, public officials who want to compare impacts in one part of the city with those elsewhere—all may rely heavily upon background indicators. The neighborhood itself may find it useful to communicate with their external audience using such indicators.

CRC defined forty-five background indicators. For instance, every neighborhood in the city compiles a profile of its population by race. These counts are seldom under the control of the community itself. Still, the racial mix in any given community is an important measure to be aware of when engaging in community activity toward sustainability. Similarly, knowing the percentage of homes that are owner-occupied may not in itself suggest what the local sustainability strategy should be, but it certainly influences strategic calculations of the first steps to take. Comparing such measures across diverse communities is often useful for at least gaining a rough sense of comparative conditions. Background indicators, however, may not be as adept at expressing linkages among various issues nor as concise in expressing sustainability concerns, but they still help fill in the picture—much as scenery helps an audience gain a better feel for the context in which stage action takes place.

A fourth category of indicator also became necessary to define as we proceeded in our initiative. Deep sustainability indicators surfaced as we brainstormed indicators for the neighborhood. Often these emerged as very pithy insights into the long-term potential of the community. Yet these indicators were impractical for current use. One example of a deep sustainability indicator is the measurement of the strength of local credit sources. A more sustainable neighborhood, we reasoned, would have its own sources of capital to promote local development. It would not solely depend upon external sources. Almost a fantasy to contemplate at the present time, this measure could potentially prove highly useful in the long term.

Deep sustainability indicators also provided a context for imagining what the community would measure after it made early progress in sustainability. As systems changed, the issues of primary importance might also change. Through practical experience, the community might discover more penetrating measures of sustainability. Defining deep sustainability indicators allowed us to keep track of ideas that promoted this sense of long-term progress. In essence, the indicators were more a visioning tool than a measuring tool. Including them in our process kept vital issues from being overlooked—even if not implemented for many years. For this reason, deep sustainability indicators were included in this discussion but are not shown in Figure 1.

Original Local Data Required

Careful combing of local databases showed that existing data sources were not sufficient for assessing neighborhood sustainability. CRC created a resident survey so Seward could compile demographic information, profile neighborhood concerns, and ascertain consumer preferences. Results from a random sample of 10 percent of neighborhood households showed that our data correlated well with federal census data, although renters and new immigrants were underrepresented in our sample despite diligent efforts. The survey not only provided solid baseline data for Seward to use in future development efforts, but also identified shops at which local residents do (and do not) shop. In addition, it provided a means for residents to convey to the neighborhood organization which issues were not being adequately addressed.

Assumptions Made by Residents

Reviewing the indicators selected by SNG shows that neighborhood residents based their selection on several assumptions. First, as a neighborhood becomes more cohesive—as residents begin to know each other better and gain more trust—local action will be increasingly effective in addressing issues like pollution, public safety, and economic development, SNG assumed. Thus, several of the most key measures assess progress toward making sure that individual residents and families develop stronger bonds with each other and more commitment to the community. The "friendly spaces" indicator, an original indicator devised by this group of Seward residents, measures the capacity of physical spaces in the neighborhood to facilitate resident connections. A second indicator assesses how many local residents shop at local businesses. Each measure invites data collection by local residents, which may include youth or other newcomers to neighborhood work, making them tools in community education and mobilization as well as in evaluation. Each represents a homegrown way to measure social capital.

Second, Seward looked more at *linkage* than at specific issues. In fact, as various technical experts reviewed this slate of indicators, one of the recurrent themes was that each person felt some key indicator was missing. One felt there were not enough measures of health; another expressed concern about the lack of education indicators. Still others complained that environmental issues were not thoroughly addressed, and so forth. Many reviewers found gaps in the treatment of their favorite issue. Time and again, residents responded that "single-issue" indicators did not express the *linkages* among issues that they had labored to define. These critics were reminded that these single issues *were already connected* to Seward's chosen indicators. The purpose of the systemic indicators was not to offer a complete list of all possible neighborhood indicators but to define linkages and focus strategic action that would most promote neighborhood sustainability. Residents opted for a shorter list of linked indicators rather than a longer list of single-issue measures.

Residents were also clear that this is simply an initial effort. Decades of future research are proposed by this slate. For instance, one can envision a day when it is possible to know how much fossil fuel energy would be saved if 10 percent more residents began to shop regularly at local stores. Over time, more precise measures of the interaction of commercial development and housing costs may be devised by professional researchers. In this sense, the residents view their work very much as a living document that will undergo revision as more expertise is gained and as conditions and issues change.

Indicators Reflect Local Goals

It should also be kept in mind that the indicators selected reflect Seward's existing strategic priorities. These are, as stated in SNG's community plan: (1) all structures will be placed in decent condition; (2) crime rates will steadily decline; (3) positive community attributes will be enhanced; (4) residential buildings will be rehabilitated; (5) a more cohesive social environment will be built; (6) a stronger local economy will be developed; (7) impacts of neighborhood life on the natural environment will be reduced; and (8) alternative forms of transportation will be promoted.

The lesson of linked indicators was not lost on SNG. In fact, the strongest outcome from this indicators initiative was Seward's decision to ensure that volunteer resident working groups, addressing the eight issues listed above, formed stronger linkages with each other. Representatives from each of these working groups now hold regular meetings to coordinate their work. Neighborhood leaders say they hope in the future to revisit neighborhood goals to define more linked goals that reflect systemic concepts of sustainability. Thus the indicators project was successful in instilling a stronger systemic approach to implementation at SNG. Due to funding cutbacks, staff turnover, and lack of a champion following the end of the granting period, the actual implementation of these indicators has not been accomplished. Still, the insights gained in this effort informed a subsequent community measurement effort in North Minneapolis.

NorthWay Community Trust (North Minneapolis)

In response to an initiative of the Northwest Area Foundation in St. Paul, which is dramatically shifting its philanthropic priorities to work in greater depth with specific low-income communities for an extended period of time, CRC also worked with resident leaders of North Minneapolis to set up an internal evaluation process. This initiative, conducted from 2000 to 2002, built upon insights gained in the sustainability indicators effort described above.

North Minneapolis includes one-quarter of the city, an area of thirteen neighborhoods containing a population of 65,000. Its southern neighborhoods have experienced some of the deepest poverty in the Minneapolis-St. Paul metropolitan area. The foundation has made a ten-year commitment to donate $10 million to a community service organization whose purpose will be to coordinate neighborhood activities to reduce poverty and build wealth among the lowest-income quartile of the community—the six thousand households earning less than $20,000 per year. CRC was invited by the foundation (with agreement of neighborhood leaders active in planning this initiative) to provide comprehensive background data covering demographic and economic conditions in the community. By mutual agreement, we took an asset-based approach in compiling this community profile: we compiled a detailed list of community assets before we studied the community's deficiencies.

Although a detailed list of community assets is beyond the scope of this essay, some highlights of our research merit mention. North Minneapolis has one of the more diverse populations in the city. The community's housing stock includes some of the most durable Victorian and pre-World War II homes in the city, and property values have been rising. Residents held $885 million in purchasing power in 1999 and pay $35 million in taxes. Workers hold a variety of administrative support, service, manufacturing, professional specialty, and executive/administrative occupations. The community supports more than four hundred businesses, from small retail to large manufacturing. There are more than 170 civic organizations in the community, linked through existing partnerships to 173 external organizations, among them the most powerful businesses and foundations in the city (New Planning Work Group 2002).

None of this obscures the very real difficulties the community faces. Residents identified six key issues: (1) residents need stronger capacity to accomplish their goals; (2) optimal land use will not be achieved without a stronger community voice; (3) neighbors lack political clout; (4) existing development efforts fail to build significant wealth for low-income people—and existing economic structures drain as much as $450 million in potential wealth away from the community annually; (5) North Minneapolis faces a substantial threat of gentrification; and (6) the northern tier of the community, home to an elderly population that raised families here and is now moving away or dying, has less citizen capacity than the southern tier, where people of color, especially, have spawned a variety of civic organizations. The hope of the NorthWay Community Trust is to build the capacities of residents to fill these gaps, knowing that this work will primarily draw upon the assets the community already holds.

Most relevant to the discussion of sustainability measures is that the residents' own analysis of poverty—similar to Seward's neighborhood analysis—is that poverty is *structural*. That is to say, powerful economic structures not only create poverty but also sustain poverty, working against citizen efforts to build wealth. The flow of $450 million out of the community each year is perhaps one of the most essential measures of this fact.

Accustomed to decades of "poverty reduction" efforts that were designed by external institutions and that reinforced the worldview that poverty was a failure of individuals, residents viewed this structural insight as key to their own poverty reduction initiative. This analysis further led them to a theory of change that assumed that addressing poverty required connecting existing residents, resources, and programs to create *systemic* change, reducing the tendency to address issues as single concerns isolated from each other. Residents sought to dismantle "silos" that isolate people and extract wealth and to build "systems" that build social connectedness and community wealth.

Under the work agreement, residents will focus their efforts in five key strategic goals: (1) building connection and capacity; (2) building community wealth; (3) creating affordable housing; (4) addressing health disparities; and (5) building knowledge. Residents assumed in this set of goals that action in all five areas will be complex and integrative and will foster systemic change. One unique aspect of the fifth strategic goal, to build a knowledge base, is that the NorthWay Community Trust expects to build such a rich collection of data covering community conditions and outcomes that the trust will become the data provider of choice. Taking this stance was important to residents who have grown weary of outside professionals launching studies of their "condition" and who then held these data close in their own files, invisible to residents and their local activity.

The trust contemplates creating an Internet-based data tracking system that will allow individual initiatives under its umbrella to post data that documents their poverty reduction and community wealth building efforts. Real-time access will allow a high level of transparency, so that any resident who chooses to learn more about program outcomes and systemic change can view measures quite readily. Hopefully, concise summaries of this data can be compiled by NorthWay staff, providing solid information to community policy discussions. External parties, such as reporters, funders, or businesspeople, who wish to learn about the community may be given access to summaries of this data, in concert with academic or administrative data. The Foundation also expects to frame its own, somewhat separate, evaluation process.

This theory of change, in turn, posed an evaluation challenge. How to evaluate this ten-year initiative in systems terms? Could we identify key indicators that showed system dynamics, rather than focusing on specific symptoms? Could we identify key "leverage points" where existing systems would be more susceptible to being changed? Could we identify integrative indicators that cut across issue "silos"?

Raising these challenges to the residents was in itself an effective strategy, since it forced our planning group to develop a more systemic analysis of poverty and a more precise theory of change. Essentially, the time frame under which an agreement between the residents and the foundation had to be signed was too restrictive to fully answer these questions. In the final contract document, only three broad outcomes were specified: (1) North Minneapolis neighborhoods are stable, healthy, and diverse; (2) poverty is reduced, and wealth is built, for the lowest quartile of the population; and (3) North Minneapolis citizens effectively act and advocate for themselves.

The agreement between the trust and the foundation rests upon a fundamental assumption that carries immense importance if accurate. If wealth is actually accumulated by low-income people in North Minneapolis, this set of indicators assumes that this will be evidence that systemic change is under way. This may serve as a reminder to the foundation, as well as to the community, that all face an uphill path against immense obstacles. This may discourage some individuals, yet there is also the potential in this initiative to understand that there will be no quick victories amidst an economy that extracts wealth from low-income communities in both urban and rural areas. The foundation's $10-million investment is intended to lay the foundation for a sustained process of systemic change and to build new relationships and new structures that make systemic change more likely—but it is unlikely to eradicate poverty, as some may have hoped at the beginning. Increasingly, the foundation speaks of investing in the community because the community itself had already reached a "tipping point" before the foundation involved itself. One scenario is that insertion of additional resources will provide the impetus for conditions to "tip." Another scenario is that the foundation's investment will create new social structures and interpersonal connections that will lay the foundation for long-term systemic change. By setting expectations to a more realistic plane, it is hoped that this evaluation process will ultimately prove inspiring rather than discouraging.

Early Thoughts on Measuring Systems

We are still in very preliminary stages of systemic indicator work. Only a few preliminary conclusions may be drawn from our experience so far. First, systemic indicators are different from unidimensional ones. Systemic indicators show linkages across issues that are often treated separately by academic disciplines, professional specialists, or citizen committees that are set up to address single causes. Unidimensional indicators tend to express more linear relationships concerning single issues. Second, with respect to system dynamics, one way to measure systemic change is to track specific linked measures that illuminate key system conditions. For instance, in North Minneapolis, one measure of the current extractive economy is the amount of potential wealth that is extracted by interest payments made by low-income home buyers, and which does not cycle back for further investment in the community. Similarly, one measure of an economy that builds community wealth is the amount of money that low-income residents are able to save over time. If this indicator emerged from an accurate assessment of the economic dynamics in a given low-income community, then one may be able to track system conditions pre- and post-transformation. Third is the issue of state-pressure-response: a related and useful paradigm from the natural sciences is to analyze the existing state of a system, measure the forces that apply pressure to that set of existing conditions, and then measure the system's response. This model recognizes that the "new" state is in turn affected by new pressures, possibly created by the system change, and leading to cycles of new states and responses. Developed for environmental systems, this may also be useful in looking at the "ecology" of neighborhood issues. Clearly this approach overlaps with the concepts mentioned above. The Sustainability Institute in Vermont has developed training tools that ask participants to identify the ways in which a system will resist change, so that this "push-back" can be measured and accounted for in the effort's theory of change.[8] Finally, there is a need to accurately identify leverage points at which a given system is vulnerable to change, which is a daunting task. This seems to depend on capturing solid insights from local residents who have worked at fostering systemic change—people who know local constraints and have experienced the prevailing system as it resists being transformed. At this stage, such analysis is definitely more of an art than a science. Still, if a group of people who represent diverse voices have established a great deal of trust with each other and can persist through complex strategic planning over time, there is every reason to expect that new insights into leverage points can be gleaned. These insights can be refined through subsequent practical work.

Common Themes Emerge

Across the diverse communities that have tackled sustainability measurements, several common themes emerge. These themes are potent despite the lack of longitudinal data that would illuminate the practical utility of these indicators.

An asset-based approach holds great power. Compiling an accurate sense of the assets a community holds, as a first step in a community improvement process, makes sense for several reasons. Building the assets a community has creates a more energizing conversation than one focused on liabilities or "problems." By recognizing that the community's answers to the issues it faces are shaped most importantly by its own capacities and assets, this places the momentum of the effort squarely in the hands of residents, where it belongs. Residents who negotiate with external partners on the basis of what they have to invest are in a better position to create a sustainable future than residents who begin by listing deficiencies.

Engaging diverse stakeholders in all phases is essential. Professional experts and technicians can only answer some of the questions that must be addressed in a sustainability effort. If technical models were sufficiently comprehensive to address complex social and natural systems, this power could grow. However, local residents have important insights for several reasons. Local residents have practical experience, often covering a long stretch of time that cannot be replicated by technical experts living outside the locale. Since local residents tend not to be specialists, they may be more likely to view a given issue from a holistic perspective rather than within the confines of an academic specialty or discipline. Groups of residents who learn from each other's diverse viewpoints and varied power interests may devise a more systemic understanding of their context. Their own efforts to change systems are likely to inform them as to leverage points at which the system may be altered—or may resist change. Intuitive and practical insights like these can be blended with technical resources—and this argues for engaging diverse resident stakeholders from the onset.

A holistic view of community and sustainability is essential if key systems are to be understood. Combining concerns of economics, equity, and ecology is one good way to do this. So is bringing all stakeholders together into respectful, transparent, strategic decision-making. Embracing complexity is essential to systemic work. We are especially indebted to the work of Glenda Eoyang (Eoyang & Berkas 1998) on complex adaptive systems for offering us insight into how systems that are themselves undergoing transformation can be effectively analyzed and affected.

Building a theory of change appears to be an effective way to build a more unified local view. Although it may prove to be a technical exercise that places high demands on the participants, devising a theory of change at the front end of a sustainability process appears to be very useful. If all affected stakeholders are engaged in the creation of a theory of change, it can lead to a far more focused effort. Consequently, if the process of defining a specific theory of change proves divisive, or if important constituencies or issues are overlooked, the entire effort could be harmed.

In North Minneapolis, issues were clarified when the group had to ask itself, "What do we mean by reducing poverty?" This question could be answered in a wide variety of ways (e.g., median income for the neighborhood rises above the poverty level, the number of welfare recipients drops to zero, or residents of color earn the same income as white residents). Any of these measures may suggest specific strategies. Any of them may be unsatisfactory (the first measure may be a sign the neighborhood has gentrified, rather than an indication that low-income people have built wealth). Most importantly, if some members of a community endeavor to measure success one way and others measure it another way, and if this difference is not made explicit, grave schisms can result in years subsequent to the process at hand.

Building a theory of change also allows residents to make explicit the assumptions they have made in their sustainability work and to check themselves on the changing nature of their context. If conditions turn out to be different than expected, or if early efforts are more effective than anticipated, having an explicit theory of change allows participants to note shifts in their approach or alterations of prevailing conditions. A theory of systemic change may well incorporate some way of assessing how the system resists change—how it pushes back against resident efforts to progress. Knowing this resistance will emerge is an effective antidote both to unreasonably high expectations of change and to the likely frustrations that result when early successes lead to greater resistance.

Our experience has been that linked indicators allowed residents to work together with more efficacy, by formally recognizing local insights and encouraging new connections among residents and new partnerships with external parties. Addressing systemic challenges inspired a more penetrating view of social, economic, and ecological challenges, giving participants a deeper understanding of their work. Residents became more engaged because they experienced a sense of vitality that had been missing in more unidimensional methods. At the same time, there are also risks in this strategy. Raising systemic concerns may also deepen tensions within a citizen effort. If there are unresolved conflicts within an organization or campaign, these certainly emerge as issues are addressed in an integrated manner. As with any intervention, evaluation work must be done with care. Still, our results show that this integrated approach holds strong promise.

Next Steps

This essay constitutes an early report that outlines preliminary findings. It is easier to hint at potential outcomes than to offer definitive insights. What is clear is that the conceptual framework developed here, in the crucible of immersed neighborhood activity using participatory research and analysis techniques, has successfully produced an increasingly comprehensive understanding of how systemic indicators of sustainability may be devised and applied. It is also clear that posing the question of systems change *in itself* may animate community action. There is little track record of practical success at this stage; nevertheless the fact that these insights have proven robust in several diverse settings on one major metropolitan area suggests that there are solid lessons to be drawn from this work.

At this writing, CRC is embarking upon a new initiative, also funded by the Minnesota Office of Environmental Assistance, to define indicators of sustainability for the city of Minneapolis. This work will build closely upon the experiences described above. This work can be referenced by viewing the training materials used in our public input process and the citizen-defined "Fifty-Year Vision and Indicators for a Sustainable Minneapolis," available at www.crcworks.org/msi.html.

Notes

1. See *Sustainable Seattle* reports, available from Metrocenter YMCA, 909 Fourth Avenue, Seattle, WA 98104, <sustsea@halcyon.com>, or <http://www.scn.org/sustainable/susthome.html>. See also Campbell (1996). Also the following Web sites: Sustainable Communities Network <http://www.sustainable.org>, Center for Excellence for Sustainable Development <http://www.sustainable.doe.gov>, Center for Sustainable Communities, University of Washington <http://weber.u.washington.edu/~common/>, Jacksonville Community Council <http://www.jcci.org>, Minnesota Planning's Minnesota Milestones <http://www.mnplan.mn.us>, Redefining Progress <http://www.rprogress.org>.

2. Maureen Hart can be reached at <mhart@tiac.net>. See also the excellent Sustainable Measures web site <http://www.sustainablemeasures.com>.

3. Comments to University of Manitoba urban design graduate student team visiting community development sites in the Twin Cities, October 14, 1998.

4. Minutes of the 1998 roundtable are available on the CRC Web site: <http://www.crcworks.org/rt1.html>, via e-mail at <kmeter@crcworks.org> or by mail at P.O. Box 7423, Minneapolis, MN 55407. The *Neighborhood Sustainability Indicators Guidebook* published for NSIP is available for free download as a PDF file from <http://www.crcworks.org/guide.pdf.>

5. The term "data poetry" had previously been used in a slightly different sense by Maureen Hart and other experts. In our usage, "data poetry" refers to indicators that have the quality of *transforming* the awareness of participants, moving action to a deeper, more systemic level. In this respect, well-chosen indicators are similar to good poems, which may transform the reader's or listener's awareness of the imagery or the subject matter interpreted by the author. Response to this term has been divided. Some professionals consider this to be a term that lacks the proper technical patina. For these settings, we now prefer the term "systemic indicator," as used in this essay. However, at the neighborhood level, we found the term "data poetry" exceedingly well received. Seward neighborhood leaders, for example, took new courage when they heard the term. The term alone seemed to transform the discussion of measuring neighborhood sustainability into a new realm—less the official province of technical experts and more a matter of local wisdom and inspiration. The term "data poetry" reinforced the sense among residents that they could exercise their own judgment effectively and create new measures that had never been invented by technical experts. This, in fact, turned out to be the case in Seward, and a working poet living in the neighborhood was ultimately one of the leaders in inventing new indicators for the community. The term, at least in this context, thus drew out wisdom and new leadership from the community that could easily have been obscured by a purely "technical" discussion of measurement. Nor was this community queasy about using the language of systems: members of our steering committee brought systems analysis texts to the second training meeting. However, not until the term "data poetry" emerged in our discussions did residents fully engage in the spirit that they could shape the measurement process to fit their local wisdom.

6. Henry Lickers, director of environmental affairs of the Mohawk Council Akwesasne, speaking at the October 1996 Sustainable Communities conference sponsored by the Minnesota Office of Environmental Assistance (OEA).

7. See *Sustainable Seattle* reports (note 1).

8. See the Sustainability Institute Web site at Since this essay was written, even more useful systems evaluation tools have been used. See, for example, the author's chapter, titled "Evaluating Farm and Food Systems" in an upcoming book to be published by the American Evaluation Association, edited by Bob Williams and Iraj Imam.

References

Besleme, Kate, and Megan Mullin. 1997. "Community Indicators and Healthy Communities." *National Civic Review* 86: 43-53.

Campbell, Scott. 1996. "Green Cities, Growing Cities, Just Cities?'" *Journal of the American Planning Association* 62: 296-312.

Eoyang, Glenda H., and Thomas H. Berkas. 1998. "Evaluation in a Complex Adaptive System." Available at <http://www.winternet.com/~eoyang/EvalinCAS.pdf>.

Kretzmann, John P., and John L. McKnight. 1995. *Building Communities from the Inside Out: A Path Toward Finding and Mobilizing a Community's Assets.* Chicago: American College Teachers Association (ACTA) Publications.

Maclaren, Virginia W. 1996. "Urban Sustainability Reporting." *Journal of the American Planning Association* 62: 184-202.

Meter, Ken. 1999. *Neighborhood Sustainability Indicators Guidebook.* Minneapolis: Crossroads Resource Center.

Wackernagel, Mathis, and William Rees. 1996. *Our Ecological Footprint.* Gabriola Island, BC: New Society Publisher.

World Commission on Environment and Development (WCED). 1987. *Our Common Future.* Oxford: Oxford University Press.

Zachary, Jill. 1995. *Sustainable Community Indicators: Guideposts for Local Planning.* Santa Barbara, CA: Community Environmental Council, Inc., and Gildea Resource Center.

The Design of the Built Environment and Social Capital
Case Study of a Coastal Town Facing Rapid Changes

Karla Caser

Introduction

In the last century the design of the built environment has faced considerable paradigm changes. The functionalist approach to design emerged in architecture and planning in the nineteenth century (Pérez-Gómez 1983), as these professions became enamored with the technological products of the Industrial Revolution. Functionality and universality of solutions were the premises of the Modernist movement, which proposed the creation of a new world for a "new technological man." Modernism prevailed for half a century, reaching its peak during the reconstruction period following World War II. In the postwar era, the first signs emerged that the new technological man did not, in practice, fit well in the new built environment based upon "universal" design principles.

In the 1960s, neighborhood activism and social movements emerged "in a struggle to preserve and enhance places that mattered" (Ley 1989: 53). In academia, an interdisciplinary approach also emerged as a way to make sense of a world that did not fit into the Modernist framework. In this way, and to counter the functionalist approach, phenomenological, anthropological, and sociological approaches came to be applied to design practices. These disciplines began to influence post-Modern design and its attendant concerns with place and with the alienation of people from their surrounding environment as a result of industrialization, urbanization, and the machine aesthetic of Modernism (Ley 1989).

Post-Modern design proposals have begun to be criticized for overemphasizing phenomenological aspects of place and not touching upon forms of power and control delivered through the built form (Berman 1982; Ley 1989). Instead of looking at place as locale, post-Modernist design has reproduced the same error of Modernism's formal emphasis by mimicking traditional built forms that are associated with ideal community life. "New Urbanism," an example of this type of design, emerged as a response to the conventions of suburbia and of more recent gated communities. New Urbanist forms, as proposed by architects Duany and Plater-Zyberk (Duany et al. 2000), are inspired by early nineteenth-century small American towns, their public spaces, and their mix of functions. The small-scale New Urbanist developments with "meaningful" public spaces were designed with the intent of promoting social interaction. However, Duncan and Lambert's (2002) analysis has shown how these developments have turned into income-segregated communities. Although New Urbanism might have solved some formal problems of suburbia, it left the inherent problem of suburbia untouched; it has dealt with the insecurity of contemporary living by seclusion into places that reify an ideal past.

The unprecedented movement of ex-urbanites and tourists in search of secluded places, namely away from big cities to rural communities, is changing the face of the countryside. Rural communities, being the foci of such "defensive" urban-to-rural migration, are in need of practices that can help them "sustain" themselves amid the rapid changes they are facing. To this end, design practice can help reduce or control the present state of "spatial confusion" (Jameson 1984) without supporting "exclusionary physical structures" (Duncan and Lambert 2002). Architectural disciplinary boundaries can also be stretched to bring in disciplines that offer a critical purview to professionals who aim to design a sustainable built environment responsive to the various constituencies within a community of place. In this essay, I explore social-spatial relationships in a coastal town facing rapid changes and, in so doing, advance Bourdieu's theory of social capital, drawing attention to its physical dimensions.

Social capital is a concept that has been widely used to explain why some communities are more successful than others at coping with changes that generate social conflicts and environmental problems. Social capital, generally described as social networks that can be mobilized for the owner's benefit, is a way to understand and deal with issues related to community sustainability (Woolcock & Narayan 2000). However, the social capital literature applies the concept to human relationships but ignores the relationship of people to the environment; the built environment is briefly mentioned only in a few studies (Federation of Calgary Communities 1999; Krishna & Shrader 1999; Tolbert et al. 2002). Alas, this downplay of the built environment occurs even though it is reasonable to argue that social capital, by being enacted through social interaction, influences and is influenced by configurations of the built environment in which these interactions take place and by its ability to help constitute group identity. According to Bourdieu (1999), social capital is objectified by proximity and openness of the built environment. Bourdieu's concept of social capital, which assumes that social capital is objectified in the built environment, emerges as an important construct to inform design practice but needs further elaboration to make explicit its connection to the physical realm. Bourdieu uses the concept of social capital quite differently from the more normative approaches of Putnam (1993, 1995) and Coleman (1990), thereby offering a critical analytical framework to issues of social-spatial relationships.

Bourdieu's (1984: 101) "praxeology" is made up of three main concepts; their relationship is schematically expressed as [(habitus) (capital)] + field = practice. For him, human action is a result of the interplay between people with different habituses and levels of capital, in various fields. Bourdieu defines "habitus"[1] as a disposition that governs action. "Field" is an arena governed by its own logic within the overall social space and resembles an "autonomous social microcosm," according to Bourdieu and Wacquant (1992: 97). Bourdieu suggests four different forms of capital: economic, cultural, social and symbolic capital;[2] all are objectified in physical space. For Bourdieu (1977: 90) the built environment can work as a medium to grasp "the structuring structures which, remaining obscure to themselves, are revealed only in the objects they structure." The relationship between field, habitus, capital, and habitat is ontological: habitus is a field and habitat is incorporated in the body; moreover, habitus is also embodied capital, and habitat helps constitute a certain habitus. Bourdieu's theory is capable of informing a critical design practice by revealing how power is crystallized in built form as objectified cultural, symbolic, and social capital. However, the physical aspect of his concept of social capital has much to gain from an incorporation of complementary theories of place; in turn, Bourdieu's praxeology offers the necessary critical purview to prevent accepting such theories unwittingly.

Methodology

This case examines the ways that the design and form of the built environment and physical space objectify social space and thus hinder or contribute to social capital formation. It does so by examining how different social groups within a small coastal town negotiate and manipulate the physical space to support their sense of place and increase their amount of social, cultural, and symbolic capital, as well as power. It focuses on objectified social capital because both objectified cultural and symbolic capital have been extensively analyzed in the literature (Philo & Kearns 1993; Duncan & Ley 1993; Dovey 1999, 2002; Duncan & Lambert 2002), whereas it has not been the case for studies of objectified social capital. This study is based upon fieldwork undertaken on an island community in the Gulf of Mexico.[3] The locale was selected with attention to three main criteria: (1) small scale, so as to make the study feasible for a single researcher; (2) social diversity,[4] reflected in the diversity in architectural housing typologies and distinguishable patterns of land use; and (3) signs of recent physical change, crucial to trigger participants' awareness of their social-spatial relationships. Port Aransas met these criteria; furthermore, the community was already involved in discussions of social-spatial relationships, more specifically about how new souvenir stores were changing the town's character and quality of life. A variety of methods were used: in-depth interviews, participant observation, and analysis of newspaper articles, aerial photographs, maps, memory boxes, old photographs, and census data. This study includes various residential cohorts based upon when they settled in the town, such as "old-timers," "mid-timers," and "newcomers," and different social classes and groups within the town.

The Setting

Port Aransas is the only settlement on Mustang Island, Texas, a barrier island on the Gulf of Mexico. It is a rich habitat for wildlife and has a variety of environments: Gulf shore, flatlands, dunes, and back bays. Originally inhabited by Indians, Port Aransas was first colonized by families of Anglo-Saxon and French origins in 1855. It has a small Hispanic population. The white settlers raised cattle and sheep, the main economic activity of the area. They were also responsible for the lighthouse, which was built in 1855 to guide the ships through the Aransas Pass (Kuehne 1973), which connects the Gulf of Mexico to the lagoon behind the island and, consequently, to the mainland. (See Figure 1.)

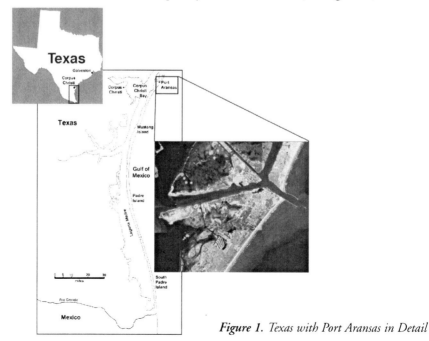

Figure 1. Texas with Port Aransas in Detail

The population at the turn of the twentieth century was composed of a few families. They formed a very close-knit community that, through intermarriage, became connected by kinship ties over the years. The houses were built out of wood from shipwrecks found on the beach; these wrecks also supplied fuel for the wood stoves. Cattle and chickens were raised as part of a subsistence economy. The relative isolation set the tone of the place. A nine-car ferry, which cost $1 and stopped running at 9 P.M., made the connection to the mainland. There were no roads to the south of the island and the beach was used as road to reach the wood causeway that connected the southern part of the island to the mainland, eighteen miles south of the township. The early twentieth century was a period of intense growth for the village, and the development of coastal transportation brought an influx of construction workers and visitors to the area. From its origins as a small village with a subsistence economy, Port Aransas would become a fishing and hunting destination. Although winter was duck hunting season, most tourism occurred during the summer, attracting elite families and single fishermen.

The town started to change after the first big wave of in-migration of the 1970s, just after Hurricane Celia made landfall in the middle of town in August 1970. Many residents use this hurricane as a reference point in time: there are people who arrived either before or after the hurricane. Most mid-timers arrived after Hurricane Celia; some of them were a part of the alternative lifestyle movement of the 1960s and early 1970s. During the 1980s, there was a relatively small population increase, around 20 percent, as the local economic boom in the oil-producing areas of Texas began to unwind when crude oil prices weakened. In the 1990s, the town increased its permanent population by 50 percent, chiefly through an influx of early retirees, but also of people attracted by its small-town atmosphere and good schools. The number of tourists also increased, as the area's geographical isolation diminished: the nine-car ferry was replaced by various twenty-seven-car ferries that run around the clock with no fare, and a highway was built that connected the island to the mainland by a concrete causeway.

These newcomer in-migrants of the 1990s, mainly early retirees, have common interests with the "old hippies" of the 1970s, namely the desire to escape the busy city life and live closer to a more natural environment. They, along with some mid-timers, are the ones who value the isolation of the island, its natural environment, its fauna, its diversity of use, and its small-town atmosphere. Since the influx of in-migrants in the 1970s, developers and business people have begun to create a year-round economy to attract a broader range of tourists, other than fishermen. Port Aransas has also become a beach destination for college students and "Winter Texans,"[5] an economically vibrant and relaxing beach resort, and, more recently, a mecca for birding. The local chamber of commerce has been actively advertising the town to encourage mass tourism and to solicit potential in-migrants and businesses. All these elements, taken together, have helped Port Aransas's population reach 3,300 people. This influx of outsiders has been as traumatic as hurricanes, especially for the old-timers.

Besides the change in lifestyle and high taxes, there are other unwelcome changes that were considered worth mentioning by old-timers: building on top of the dunes and the wetlands, city zoning and regulations, privatization of the waterfront, relocation of the post office, parking fees on the beach, and beach scraping and grading. Not only old-timers complain about these changes; some mid-timers, especially the ones not involved with real estate or business, are quite vocal about the detrimental effects these have had on everyday life and the natural environment. Community change always brings forth discussions about whose view should prevail. In 2001, the arrival of new chain souvenir stores onto the public scene inspired discussions about the importance of the built and natural environments to the quality of life, identity, and sustainability of the community. In 2002, the New Urbanism firm of Duany and Plater-Zyberk designed a planned community of 2,500 houses and two golf courses, likewise inspiring debates about the development and its impact on community life.

Habitus and Habitat

Many are the ways in which habitat helps constitute Port Aransas residents' sense of place, one that makes for a certain type of residential attachment and a distinct "habitus." Perhaps foremost is its geographical isolation from the mainland; being located on an island is paramount for enacting a sense of place. Residents talk about "us" and "them," namely the mainlanders, similar to many other islanders depicted in studies and novels (Hansen 1993; Harris 2002). Islands imply both symbolic and geographical isolation, and in Port Aransas, this isolation is reenacted in daily trips on the ferry. The ferry crossing is another way habitat enacts the town's sense of isolation. The ferry is recurrent in all interviews, signaling its significance, as if it is a ritual undergone to reach *freedom.*[6] The sense of freedom is often perceived as being physically objectified in Port Aransas; according to one resident, "when you hit the ferry you know you are here …and when you hit the land you know you hit freedom!" It seems that the ferry also works as a form of catharsis, a moment of release from tensions, not only because it is part of the ritual of reaching the island but also because you can feel you are "not in control" of the car. The car is a well-known symbol of one's life, and daily trips on the ferry, together with the expectation of severe tropical storms that no one is able to control, free residents from feelings of "omnipotence" with respect to nature and its boundary conditions. In this way, habitat appears to help constitute habitus, through displacements of the body.

If islands are well known for their mythical sense of place, Port Aransas embodies this exponentially, for it is a barrier island, described by Hansen as more "process than substance" (1993). Strong winds, high humidity, isolation, and scarce vegetation all help to enact Port Aransas's sense of place, one of intense physical changes and impermanence that reverberate in the built environment, for example, *funky buildings* and houses on pilings, and on people's habituses. It seems that these physical characteristics and the town's relative isolation set the tone of the place, enacting dispositions/habituses and attracting people with similar dispositions or habituses, which in turn contribute to its overall sense of place. An example of habitus/habitat relationship is the way most of the newcomers and mid-timers clearly articulate how habitat helps constitute identity, one that involves the choice of living a simpler life on an isolated island. The "shared habitus" based on common assumptions about the natural environment and *freedom* fits squarely within North America's frontier heritage and attendant individualism.

The "escapism" (Tuan 1998) inherent in these islanders' attitudes has much in common with other cultures and places, such as the escape to rural villas in Renaissance Italy and contemporary urbanites' escape from inner-city cores to the suburbs. Historically escapism also seems to be one of the reasons for the first influx to the island in the 1850s. In Port Aransas, contemporary styles of escapism include "ritualized frontiermanship" (Smith 2002: 31) in the form of beach cruising and ritualized trips to vacation places (Gillis 2001) as in the recent tourism boom on the island. Escapism is a generalized reaction of contemporary society to modernity's conquest of time and space and its power in eradicating spatial and temporal boundaries. In this regard, Tuan (1998) argues that a strategy of escaping involves embracing not only simple geographical and physical forms but also a simpler social life. To use Bourdieu's terms, social and physical spaces tend to coexist and reinforce each other.

A common feature of the above-mentioned shared habitus is that of *living on island time,* which is simultaneously a spatial and temporal strategy, most notably for mid-timers and newcomers. As a temporal strategy, *living on island time* is a reconquest of time. Port Aransas offers a feeling of transcendence of the temporal flux, and most residents mention how they perceive themselves transcending such a flux by *living on island time*. This strategic reconquest of time is accomplished through a series of stances, both emotional and physical. *Living on island time* implies a conscious decision to scale down, to lead a simpler life. It seems to be a different type of escapism, one that peels off the derogatory meaning associated with the term, namely the escape into a secluded or fantasy world. Escaping to Port Aransas does not necessarily imply a retreat to *"paradise"* (Silver 2001), although some might expect that to be the case and either be disappointed and leave or start to change it in order to make it look more like one's idea of paradise. Escaping to the island might simply mean that the disposition of some individuals has led them to question the societal *definition of success* because they feel they need to *slow down* and spend more time doing whatever they like. It can also be argued that it is not a coincidence for mid-timers and newcomers to adopt the strategy of *living a simpler life* as a way of reconquering time. As Tuan (1998: 25) reminds us, "the perceived quality of life of peasant life–or any people who has a natural resource based living–is its timelessness." Hurricanes seem also to help this reconquest of time. Time is organized around them, as can be seen in most of the interviewees' comments, such as: people started to come after Hurricane Celia, namely in the 1970s; somebody arrived after Hurricane Carla, or in the 1960s. In a sense, they procure a "cyclical rebirth," and in the linking of people's arrival on the island with the occurrence of natural hazards, the cyclical time of pre-Modern societies is somehow recreated for contemporary Port Aransas's residents.

As a spatial strategy, *living on island time* is a reconquest of an intimate human-environment relationship. The phrase implies a "sense of oneness" (Tuan 1998), a unitary and easily apprehended background that helps set up clearly defined "cognitive maps" (Jameson 1984). It also implies the presence of clearly defined boundaries. Newcomers and mid-timers alike have come from cities with their characteristic isolation. They have taken refuge to a place that, because of its geographical isolation, reinforces their psychological isolation from mainstream American society, and in a sense, the geographical barrier helps create a distinct cultural group. As many residents mentioned, "we need to cross a barrier to get here." It has also to do with the sight of an encompassing surrounding (*water is everywhere*) and the feeling of a calming natural environment, one that a person can take in from a single vantage point and hear, smell, and taste as well. All together–sight (*water everywhere*), smell (*ocean breeze, salt*), sound (*white sound*–waves–and *whispering sands*), and touch (*ocean breeze*)–the environment in Port Aransas is the equivalent of the "pictures on the wall" of Tuan's (1998) family room, a unitary background that not only renders the built environment "legible" (Lynch 1960) but also creates a comforting sense of "oneness" (Tuan 1998). Indeed, mid-timers and newcomers alike identify *ocean breezes* and various sounds as being responsible for them *living on island time*. People appreciate Port Aransas as a place that affects their senses, which is the opposite of a Modern built environment that lacks sensory stimuli other than the visual, as Tuan (1977, 1995) reminds us.

The sense of belonging to Port Aransas appears to create a shared identity that contributes to a mutual understanding among the different groups in town. It may be because physical threats to the sense of belonging are especially relevant for a community that has its identity based on outdoor activities and scenic views. These threats pull different groups together toward a common objective. This is not to say, however, that different groups do not have different habituses according to their position in the social space. There are residents of different class, gender, and ethnic groups holding different views. The conflicting habituses and different levels of social capital determine not only what is mostly at stake in preserving their identity but also how the physical environment has been shaped. This leads to Bourdieu's concepts of field and capital and how they are intertwined with habitat.

Habitus in Conflict: Field and Habitat

In Port Aransas, a collective habitus based upon isolation and the threat of natural hazards makes for a shared sense of place, but there are divergences based on one's position in the field of power. There are conflicting habituses, as well, because different groups hold different "structures of feelings" (Williams, in Agnew 1993: 263), according to their background and position in the social space. The town's social diversity is represented by some remaining old-timers, local transients, a few working-class people still able to afford to live on the island, school and administrative employees, small business people, a local economic elite, and the upper-class retirees. The same residents that portray Port Aransas as a friendly place also complain about the social distance between the "haves and have-nots." How to explain this contradiction other than to refer to field and different habituses? Gender and class differences clearly play a significant role in defining one's position in Port Aransas's social space. For example, among working-class residents, it is the women who seem to experience the least favorable working conditions. A related question is "How and for what reasons is the town's shared habitus sustained in the face of such conflicting views?" Indeed, Duncan and Lambert's (2002) argument is that "landscape taste" can be used to obscure conflicting class relations, which would translate into hidden class conflicts obscured by the glorification of a shared island-habitus and its respective built form, designed expressly for the local taste.

Bourdieu (1999) warns that class conflicts tend to be obscured by being translated into the physical space and built environment, where it is naturalized and becomes reified social space. In a sense, the fast pace of change in Port Aransas appears to be restraining the new physical space from being taken for granted. Class conflicts have been abruptly translated into the built form as strategies for social control and, as a result, are more transparent to residents targeted by these strategies. Moreover, debates about the town's future among the various resident constituencies have been surfacing as these social control mechanisms were steadily put in place, at both physical and symbolic levels, by the town's elites.

In Port Aransas, social class distinctions are present and objectified in middle-class "aestheticized" zoning and regulations and the institutionalization of public open spaces. Examples of this abound all over town. One of them is the "beautification" encouraged by the Garden Club and by regulations and zoning. It seems that people are attracted to the town's "quirky environment," but as soon as they move in they share in a continuous effort to create a manicured quirky environment or even restrain the quirkiness to specific places. Pavement, sidewalks, lighting on the beach, parks, and birding centers are all part of this process; all are improvements required by newcomers who call for services similar to the ones they had where they came from. Being the local power brokers, businessmen, developers, and real estate agents have the power to legitimize their views, which are usually based upon a landscape aesthetic held by elites.

Not surprisingly, the elite groups' agenda creates considerable resentment among working-class residents. The ones outside the field of power have been struggling to pay increasing rent and property taxes. As described by many residents, the social space of the community has changed drastically, with many people being "washed out" of town. There is the perception that elite aesthetic concerns overlook low-income necessities with respect to the built environment, and some even see a "hidden agenda" to exclude them from Port Aransas. Other residents complain that the elites prefer to destroy nature to sell an image. The destruction of nature refers to beach grading. Port Aransas' beach area has considerable seaweed, which smells and defiles the image of an "ideal beach." The city thus took the initiative to remove the marine algae, which are a natural part of the Gulf shore ecosystem.

There are conflicting views with regard to the new souvenir stores as well. While a few members of the business community approve of them, others are concerned that the buildings will work as an incentive to mass recreation and thus devalue their real estate and the overall quality of life. Many others are concerned that the presence of these shops indicates that the local business elite is selling Port Aransas as a commercial destination, with no concern for the environment. Working people fear that the mass consumerism wrought by the tourist shops will lead eventually to the beach being closed to vehicles, with fees being charged to control entrance, which would mean the loss of a much-appreciated public space. Ironically, developers and the business community have appropriated the environmental discourse around commercialization to accuse local working-class residents of having exclusionary intentions.

While debates over commercial interests are openly voiced in municipal arenas, conflicts over civic tolerance prevail under the surface. Most middle- and upper-class Anglos view the town as a place with high tolerance for difference, a perception that is not shared by all residents, particularly those outside the field of power. In one resident's words:

> There is a lot of discrimination with Hispanics …and it is very well hidden …here you would ask the majority of the white people and they would say: we don't have a racial problem …there is a racial problem here. …things [you] thought had died a long time ago are still here …

In a sense, class and cultural conflicts are surfacing in Port Aransas out of diverse residents' attempts to define the "sense of place." It could be argued that the prevailing sense of place is still being negotiated by the various social groups in town, fighting over whose view should prevail.[7] There is not only the "small fishing village" sense of place. Others describe it as a "sport fishing resort." This view is reflected in the habitat by the adoption of a "New England style," characterized by the newly renovated chamber of commerce building at the entrance of town.

Capital and Habitat

Port Aransas offers various examples of how different forms of capital are objectified in the built environment. One form uses the built environment to secure symbolic capital through a prestigious address and type of residence. In the case of Port Aransas, it was mentioned that the mere fact of living on the island grants one a certain prestige. The island location, as distinct from the mainland, also objectifies cultural and social capital (Bourdieu 1984): low-income residents are quite aware that the geographical location may hinder their cultural and social capital. One resident states:

> They don't want the working-class people to live here ... they want us to live across the way, want us to commute ... and [have] bad schools and gangs ... over here there isn't any of that! But it is expensive!

To live outside the island deprives one not only of good schools and police but also of the possibility of augmenting one's social capital through, in Bourdieu's (1999: 127) words, "meetings at once fortuitous and foreseeable that come from frequenting well-frequented sites."

Viewing the ways elites have appropriated Port Aransas's shared sense of place affords insights into how cultural capital is objectified in those sites that are used to "sell" the town to tourists and immigrants. These sites thereby objectify cultural resources that are, as Kearns and Philo (1993: ix) point out, "mobilized by city managers for capital gain." Local authorities and businessmen draw upon the island's economic and social history to market the town, and they also profit from the increased property values and number of consumers. It is such a pervasive phenomenon that even the town's rowdiness and "funky" buildings are transformed into a marketable sense of place. Bourdieu argues that one's cultural position can be objectified in the built environment. On the island, for example, a house's architectural style is perceived as a reflection of the owner's taste and culture. A concrete example of this is the "New England style," adopted across North America precisely because it conveys the refinement associated with New Englanders. Cultural capital is currently objectified in Port Aransas's mansions on treated pilings, upper-class suburban houses, and New Urbanism architect-designed New England-style houses, all standing, in Duncan and

Lambert's (2002) estimation, as "signifiers" of Anglo upper-class "good taste." Alternatively, trailers remain as signifiers of a working-class aesthetic, an eyesore to some elite town members.

The objectification of cultural capital is not restricted to the architectural style of the building alone but extends to its surrounding environment, namely the garden. A vivid example of this link between the house environment and cultural capital is the educational purpose inherent in the island's Garden Club Association. Working-class residents, deprived of economic capital, have to adhere to a middle-class aesthetic sanctioned by the club and the town elites, for objectifying one's newly acquired cultural capital on the front lawn yields "objectified symbolic capital" for a middle- and upper-middle-class community keen on displaying such acquisitions. According to Bourdieu, objectified cultural capital can be turned into objectified symbolic capital once it is recognized and legitimized as a symbol of good taste and the distinction of the owner. The Garden Club's recognition initiative provides those households with exemplary lawns and gardens the opportunity to be featured in the local newspaper as a model of good taste and care for the community. To be short-listed as the "Garden of the Month" and to have photographs of both the garden and its owner in the local newspaper[8] are evidence of the prestige that can be acquired with a manicured garden. Various cultural geographical studies argue that in open societies with high spatial mobility, such as North America, the house, and its façade, address, and interiors–to which I add the lawn–are the primary source of status display (Pratt 1982; Duncan 1982; Agnew 1982).

On the island, symbolic capital is objectified in type of residence and neighborhood. The waterfront is by far the most prestigious and grants its inhabitants a certain distinction. Inversely, some upper-middle-class residents stigmatize trailer parks and their inhabitants, in what Bourdieu (1999) calls the "ghetto effect." Employing a design professional is another way to legitimize built forms and grant symbolic value to structures (Crilley 1993; Sandercock 2002). In Port Aransas, architects are responsible for designing upper-class vacation mansions using the manicured cottage style typical of the town. This helps to confer upon the traditional cottage style a value and recognition that it did not have before. As an example of the power bestowed upon architects, one resident describes the way in which they create objectified symbolic capital:

> it is so funny because most of the people that live in town at my age, we came to town and we lived in houses that were on pilings, square box houses with wood siding and metal roof! As we got older and more financially available we upgraded our houses to more modern and nicer homes …and then they come back in and say: "this is the thing to have!" We should've just stayed in the old houses! We sold our houses too soon!

Residential size is also symbolic in Port Aransas's "Protestant Ethic-type" society, where economic capital objectified in large estates becomes objectified symbolic capital, granting its owner a "proximity to God" (Smith 1993). The chief difference between estate-style homes on the mainland and vacation homes on Mustang Island and other coastal areas occurs in materials. Architects have recently legitimized the use of vernacular materials; as a consequence, a vacation home in granite and marble would be a signifier of the owner's bad taste, similar to wearing the wrong type of clothes and shoes to play golf.

For Bourdieu (1999), the built environment can constrain or support social capital through "physical objectification," whereby physical arrangements can enhance or deter social interactions. However, physical space is not sufficient; to fully use a habitat, there will also need to be the appropriate habitus.[9] The built environment might constrain certain residents' ability to enhance social capital through social interaction by presenting them with unfamiliar situations, which Bourdieu (1999) calls "symbolic objectification." Each type of objectification, then, can support either "inclusive" social capital when a site is conducive to social interaction between various groups, or "exclusionary" social capital when it works as a signifier of a single group or habitus.[10] In Port Aransas, there are examples of both types of objectification. Beach parking and pier entrance fees have been recently put in place as another example of "culture turned into capital" by local elites. As a consequence of the fees, some residents mentioned that they stopped going there, being forced to abandon a once-important public meeting place. In this way, such sites have come to objectify exclusionary social capital. Besides fees, the physical privatization and consequent closure of spaces along the harbor were also cited by working-class residents as another example of an initiative that is restraining their use of space and, consequently, their ability to interact with people and augment their social capital.

The post office is another example of objectified exclusionary social capital. A recent change in the internal layout of the post office, and its relocation to the outskirts of town, was unanimously perceived as detrimental to social exchange. Location affected the daily routine of going to the post office to check mail and meet people. In addition, changes in the layout seem to have created a "sociofugal" space out of a previous "sociopetal"[11] one. As a resident points out:

> I would say the new *post-office* for me ...just doesn't have the...appeal, socially ...you still say hello to people, but you don't see quite as much of that socializing going on ...it seems to be a sterile environment, I can't explain it! In the old post-office we had a board with all the outdated notes ...and this one has white walls, the floors are immaculate.

Another place that objectifies exclusionary social capital is the recent community within the city limits planned on New Urbanist principles. Despite the neo-traditional ideology, which advocates the creation of a more livable and diverse social environment, the way New Urbanist communities have been developed tends to reinforce segregation and social division while maintaining hegemonic class and ethnic structures (Duncan & Lambert 2002). Such enclaves tend to constrain the formation of social capital by lower status groups. In Port Aransas, this type of planned community is quite likely to produce an increase in the tax rate, thereby excluding more low-income people from the town, which has been increasing in frequency since the 1970s. In addition, the New Urbanist community "package" includes two golf courses. Hemingway (1999) discusses this relation between type of leisure and creation of social capital. Golf is well known as an elite sport, and the consequences of using extensive areas on an island for golf courses will clearly affect certain groups' use of this space for interaction, thereby objectifying exclusionary social capital.

Alternatively, the beach in Port Aransas still offers itself as an affordable leisure place that supports inclusive social capital. The beach helps to mitigate people's social defenses, that is, the signs of their cultural and economic capital, and helps promote interaction between people. As one resident says, "Here people are who they are …when you go down to the beach and everybody is lying in their bathing suits, you can't be anybody else than who you are!" Beach admission is still without cost in Port Aransas, and beach accessibility seems to epitomize a type of freedom that, among other things, allows social diversity. Indeed, it is generally in such public locations that "democratic" leisure activities take place. They provide the working class with an opportunity to interact with other groups on a more equal level. Tuan (1977: 117) has pointed to the inclusiveness of beach activities: "Swimming, unlike many competitive sports, minimizes the physical and social differences among human beings …It requires no expensive equipment." Waterfront areas, such as the harbor and the beach, are key places for enacting a collective habitus, but they also tend to support the generation of inclusive social capital between different groups.

Examples of symbolic objectification are also found in Port Aransas. Old-timers especially are able to perceive how an increasing "aestheticized" public space can cause a symbolic, but nevertheless efficient, privatization of the public space.[12] This aestheticized public space started to emerge with the migration of upper-middle- and upper-class residents into a once proxemic space (Hall 1966; Wallin 1998; Greenbie 1998).[13] Before this time, Port Aransas was a community connected by kinship ties, with communal activities, including dances on the streets; today it is a place with several cross-cutting class habituses and fields. Regulations emerged as a natural consequence. However, the local field is one of power as created by economic and cultural elites, and working-class people lacking the necessary "proper" taste and other class-based dispositions tend to feel symbolically excluded from public spaces.

In Port Aransas, built forms objectify social capital by serving as signifiers of one's participation in a group, class, or field. The new neighborhoods, each with its architectonic style, stand as signifiers of a "higher up" in income and more secluded group. Downtown residences stand as a signifier of a more "down to earth" and socially diverse group. Conversely, trailer parks are the areas of town overtly stigmatized by some wealthier interviewees; these built forms can preclude residents from augmenting their social capital through the stigma associated with living in this neighborhood. They can be said to objectify social capital; the built environment, as a symbol of one's alignment with a certain group, can thus support or hinder social capital. Architectural diversity seems to objectify inclusive social capital. In general, some perceive the new neighborhoods as fragmenting the spatial fabric of town. On the other hand, diversity in size and style of houses in the same neighborhood conveys a sense of oneness, a proximity in space that seems to spill over into social space, thus contributing to social capital formation among social groups; in the words of one resident, " rich people built palatial homes in town; and it gave the town the sense that they were growing together."

Conclusion

In the above case description of Port Aransas, a small community with high amenity value, we have seen how and why environmental changes affect social capital and thus contribute to social disruption within the town. Most importantly, these dynamics have clearly influenced subjective meanings of landscape value for ordinary people in Port Aransas, because the town's built environment has helped define and enact it as a "place" for its inhabitants. Similarly, rapid changes, namely in-migration, tourism, and, foremost, new developments in town, have triggered overt negotiations over place. Among the various groups in town, with their differing views regarding the community as a "place," we have seen how habitat has been manipulated by each of them to reflect and enact their position in the local sphere of power. In this regard, the study has shown the ways by which capital in its various forms—cultural, social, and symbolic—is objectified in the built environment, thus contributing to the consolidation of each group's power. Finally, this case points out four ways by which the built environment objectifies social capital: it symbolically and physically constrains social interaction and engenders identity and predictability.

This essay began with the premise that Bourdieu's praxeology could work as a framework to guide studies of place. Using Port Aransas as a case, it was possible to describe and explain how the three dimensions of place together define the town as a "place" for its residents. The negotiations over place were analyzed to enhance Bourdieu's theory of social capital, with an emphasis on the various forms by which social capital is objectified. In this light, objectified/physical-social capital can assume two forms: *habitation* and *persistence*. Habitation refers both to physical and symbolic distances and to openness of the built environment that can affect social interaction. Persistence refers to environmental cues that help individuals gain existential meaningfulness, which would enable social interaction. These cues can take two forms. One is through a unitary background; a nonfragmented town fabric conveys a sense of oneness and objectifies inclusive social capital for its residents, that is, "predictability." The other form is the built environment, which objectifies social capital as a signifier of one's position in the field of power, that is, "sign" (see Caser 2004).

This study has implications for sustainable planning and design of small communities and the management of community and landscape changes. By considering both the physical and symbolic constraints to social interaction, designers can identify existing pockets of physical-social capital that help to sustain community life. Similarly, by applying the concept of persistence, designers can find ways to nurture ontological security through the creation of meaningful built environments and thereby help alleviate some of the alienation inherent in late Modern social life. This study provides a way to consider the hidden forces objectified in the built environment and enables the designer to use the strength of the spatial configuration to build meaningful and inclusive communities of place, thus helping to create a habitat that enables more inclusive social capital to thrive.

Notes

1. Recently, the concept of "habitus" has been introduced in disciplines such as architecture and geography, to account for social/spatial relationships that help explain practices related to the built environment (See Edward Casey 2001; Podmore, 1998). In these studies, habitus is taken as "an embodied sense of place" (Hillier & Rooksby, 2002:5), a sense of place in the physical space.

2. Of the many forms of capital, economic capital is the form most often associated with the word "capital." It refers to monetary and physical resources and to material wealth in general (see Bourdieu & Wacquant 1992; Everett, 2002). The other three forms of capital are nonmaterial. Cultural capital refers to people's knowledge and skills (see Bourdieu 1986). In Bourdieu's own words, social capital refers to "the aggregate of the actual and potential resources which are linked to possession of a durable network of more or less institutionalized relationships of mutual acquaintance and recognition" (ibid.: 248-49). The fourth form of capital, symbolic capital, is the form assumed by the three other forms of capital when they are regarded as legitimate.

3. This paper is based on the Ph.D. thesis I developed at the University of Guelph, as a Capes Fellow Student (see Caser, 2004).

4. It is well known that architectural housing typologies and distinguishable patterns of land use reflect the presence of both different socioeconomic groups and habituses (see Pierre Bourdieu 1999; Paquette & Domon 2001; Abu-Gazzeh 1999) as well as changes in demographics and social structure, which in coastal areas are associated with tourism and in-migration of retirees (Wood & Handley 2001)

5. Initially called "snow birds," the northern tourists that go south during the winter season to escape the rigors of cold temperatures and snow are now called "Winter Texans," which is a more friendly term. The residents seem to perceive the nickname "snowbirds" as a little derogatory, whereas it seems that by calling them "Winter Texans," there is an explicit acceptance of their presence, their acknowledgment as Texans, even if only during winter time.

6. I use words in italics to indicate that these are concepts abstracted from interviews with residents. Often these words are recurrent in various interviews. When italics are enclosed between quotation marks (*"no no"*), it means that the concept is present in both interviews and in the literature. In this case, the reference is given in sequence. Concepts enclosed within quotation marks (" ") without references refer to Bourdieu's concepts.

7. Debates abound in public citizen meetings and in letters to the local newspaper. However, it is mainly the dominant field of power that is engaged in the discussion, with a few exceptions of the working class. As already described in studies of counterurbanization, it is the newcomer with higher education who usually gets involved in initiatives to protect what s/he perceives to be the historical past and the "character" of the place, such as the creation of architectural ordinances to control growth and a historical association in Port Aransas. Indeed, as one resident argued, what tends to be created is a "cardboard past" (see Nelson 1997; Smith & Krannich 2000).

8. It does not go without note that such awards mirror a North American moralist, "Protestant ethic" ideology that values work toward the betterment of the community–in this case in the garden–and legitimizes working citizens as outstanding.

9. As dispositions, group and class habituses tend to restrain one from certain situations, opportunities, and places. In Bourdieu's terms, the sense of "being at home" in a place "arises from the quasi-perfect coincidence between habitus and habitat, between the schemes of mythic vision of the world and the structure of domestic space" (Bourdieu 2000: 147).

10. I use a neutral term, *inclusive*, to refer to bridging social capital, a term introduced by Putnam 1995) in response to criticism levelled at his unilateral view of the beneficial nature of social capital. Putnam introduced "bridging social capital," referring to social capital generated among individuals from different groups, to counterpoint "bonding social capital," referring to social capital that is beneficial to a group and prejudicial to the ones it excludes. Bonding social capital I call "exclusionary social capital."

11. Based on Osmond (1959) and (Hall 1966: 108) describes as sociofugal spaces that "tend to keep people apart" and sociopetal spaces as these "that tend to bring people together."

12. In a similar way, Philo and Kearns (1993: 12) describe the transformation of the Parisian boulevards into "bourgeois interiors" and argue that historically, city regulations are a form of social discipline.

13. The term "proxemic" was coined by anthropologist Edward Hall (1966) to designate "the study of the symbolic and communicative role in a culture of spatial arrangements (*Random House Webster's Unabridged Dictionary* 2003) and was later developed into a pair of related concepts, proxemic and distemic, to designate the way different spaces are used by groups. Proxemic spaces would be places that are used by people with similar habitus and could also refer to a specific field. Distemic refers to places used by various groups, each with its own norms and values, requiring regulations to stipulate the proper use of such space. However, as noted in the previous section, these regulations are dictated by the local field of power according to its necessities and values, and they require a proper habitus to fully use the space.

References

Abu-Gazzeh, Tawfiq M. 1999. "Housing Layout, Social Interaction, and the Place of Contact in Abu-Nuseir, Jordan." *Journal of Environmental Psychology* 19: 41-73.

Agnew, John. 1982. "Home Ownership and Identity in Capitalist Societies." In *Housing and Identity*, ed. James S. Duncan. New York: Holmes and Meier Publishers, Inc.

___. 1993. "Representing Space: Space, Scale, and Culture in Social Science." In
    Place/Culture/Representation, ed. James Duncan and David Ley. London and New York:
    Routledge.

Berman, Marshall. 1982. *All That Is Solid Melts into Air: The Experience of Modernity.* New
    York: Simon and Schuster.

Bourdieu, Pierre. 1977. *Outline of a Theory of Practice.* Cambridge: Cambridge
University Press.

___. 1983. "The Field of Cultural Production, or the Economic World Reversed." *Poetics*
    12: 311-356.

___. 1984. *Distinction: A Social Critique of the Judgement of Taste.* Cambridge, MA: Harvard
    University Press.

___. 1986. "The Forms of Capital." In *Handbook of Theory and Research for the Sociology of
    Education,* ed. John G. Richardson. New York: Greenwood Press.

___. 1989. "Social Space and Symbolic Power." *Sociological Theory* 7 (1) 14-25.

___. 1999. "Site Effects." In *The Weight of the World: Social Suffering in Contemporary
    Societies,* ed. Pierre Bourdieu and Alain Accardo. Stanford, CA: Stanford University
    Press.

___. 2000. *Pascalian Meditations.* Stanford, CA: Stanford University Press.

Bourdieu, Pierre, and Loic Wacquant. 1992. *An Invitation to a Reflexive Sociology.* Chicago:
    University of Chicago Press.

Caser, Karla. 2004. "Physical-Social Capital: Towards a Critical Design Praxis for
    Communities of Place." Ph.D. dissertation, University of Guelph.

Casey, Edward. 2001. "Between Geography and Philosophy: What Does It Mean to be in
    the Place-World?" *Annals of the Association of American Geographers* 91 (4): 683-93.

Coleman, James S. 1990. *Foundations of Social Theory.* Cambridge, MA: Belknap Press /
    Harvard University Press.

Cortese, Charles F. 1982. "The Impacts of Rapid Growth on Local Organizations and
    Community Services." In *Coping with Rapid Growth in Rural Communities,* ed. Bruce A.
    Weber and Robert E. Howell. Boulder: CO: Westview Press.

Crilley, Darrell. 1993. "Architecture as Advertising: Constructing the Image of
    Redevelopment." In *Selling Places: The City as Cultural Capital, Past and Present,* ed.
    Gerry Kearns and Chris Philo. New York: Pergamon Press.

Dovey, Kim. 1999. *Framing Places: Mediating Power in Built Form.* London: Routledge.

___. 2002. "The Silent Complicity of Architecture." In *Habitus:A Sense of Place,* ed. Jean
    Hillier and Emma Rooskby. Burlington, VT: Ashgate.

Duany, Andres, Elizabeth and Jeff Speck, 2000. *Suburban Nation: The Rise of Sprawl and
    the Decline of the American Dream.* New York: North Point Press.

Duncan, James S. ed. 1982. *Housing and Identity.* New York: Holmes and Meier.

Duncan, James, and David Lambert. 2002. "Landscape, Aesthetics, and Power." In
    American Space/American Place: Geographies of the Contemporary United States, ed. John
    A. Agnew and Jonathan M. Smith. Edinburgh: Edinburgh University Press.

Duncan, James, and David Ley, eds. 1993. *Place/Culture/Representation.* London and New
    York: Routledge.

Everett, Jeffrey. 2002. "Organizational Research and the Praxeology of Pierre Bourdieu."
    Organizational Research Methods 5 (1): 56-80.

Federation of Calgary Communities. 1999. *The Ideal Community: Design Criteria That Foster Participation in Civic and Community Life.* Calgary.

Gillis, John R. 2001. "Places Remote and Islanded." *Michigan Quarterly Review* 40 (1): 39-58.

Greenbie, Barrie B. 1998. *Space and Spirit in Modern Japan.* New Haven, CT: Yale University Press.

Hall, Edward T. 1966. *The Hidden Dimension.* New York: Anchor Books.

Hansen, Gunnar. 1993. *Islands at the Edge of Time: A Journey to America's Barrier Islands.* Washington, DC: Island Press.

Harris, Joanne. 2002. *Coastliners.* Toronto: Doubleday.

Hemingway, John L. 1999. "Leisure, Social Capital, and Democratic Citizenship." *Journal of Leisure Research* 31 (2): 150-65.

Hillier, Jean and Emma Rooksby. 2002. "Introduction." In *Habitus: A Sense of Place,* ed. Jean Hillier and Emma Rooskby. Aldershot: Ashgate.

Jameson, Fredric. 1984. "Postmodernism, or the Cultural Logic of Late Capitalism." *New Left Review* 146: 53-92.

Kearns, Gerry, and Chris Philo, eds. 1993. *Selling Places: The City as Cultural Capital, Past and Present.* New York: Pergamon Press.

Krishna, Anirudh, and Elizabeth Shrader. 1999. *Social Capital Assessment Tool.* Washington: The World Bank, Conference on Social Capital and Poverty Reduction.

Kuehne, Cyril Matthew. 1973. *Hurricane Junction: A History of Port Aransas.* San Antonio, TX: St. Mary's University.

Ley, David. 1989. "Modernism, Post-modernism and the Struggle for Place." In *The Power of Place,* ed. James A. Agnew and James S. Duncan. Boston: Unwin Hyman.

Lynch, Kevin. 1960. *The Image of the City.* Cambridge, MA: Technology Press / Harvard University Press.

Nelson, Peter B. 1997. "Migration, Sources of Income, and Community Change in the Nonmetropolitan Northwest." *Professional Geographer* 49 (4): 418-30.

Osmond, Humphrey. 1959. "The Relationship Between Architect and Psychiatrist." In *Psychiatric Architecture,* ed. Charles E. Goshen. Washington. DC: American Psychiatric Association.

Paquette, Sylvain, and Gerald Domon. 2001. "Rural Domestic Landscape Changes: A Survey of the Residential Practices of Local and Migrant Populations." *Landscape Research* 26 (4): 367-95.

Pérez-Gómez, Alberto. 1983. *Architecture and the Crisis of Modern Science.* Cambridge, MA: MIT Press.

Philo, Chris, and Gerry Kearns. 1993. "Culture, History, Capital: A Critical Introduction to the Selling of Places." In *Selling Places: The City as Cultural Capital, Past and Present,* ed. Gerry Kearns and Chris Philo. New York: Pergamon Press.

Podmore, Julie. 1998. "(Re)Reading the 'Loft Living' *Habitus* in Montréal's Inner City." *International Journal of Urban and Regional Research* 22 (2): 283-302.

Pratt, Geraldine. 1982. "The House as an Expression of Social Worlds." In *Housing and Identity,* ed. James S. Duncan. New York: Holmes and Meier.

Putnam, Robert. 1993. "The Prosperous Community–Social Capital and Public Life." *American Prospect* (Spring): 27-40. .

___. 1995. "Bowling Alone: America's Declining Social Capital." *Journal of Democracy* 6 (1): 65-78.

Relph, E. C. 1976. *Place and Placelessness.* London: Pion Limited.

Sack, Robert David. 1980. *Conceptions of Space in Social Thought.* Hong Kong: Macmillan.

Sandercock, Leonie. 2002. "Difference, Fear, and Habitus: A Political Economy of Urban Fears." In *Habitus: A Sense of Place,* ed. Jean Hillier and Emma Rooskby. Aldershot: Ashgate.

Silver, Kenneth E. 2001. *Making Paradise: Art, Modernity, and the Myth of the French Riviera.* Cambridge, MA: The MIT Press.

Smith, Jonathan. 2002. "The Place of Nature." In *American Space/American Place: Geographies of the Contemporary United States,* ed. John A. Agnew and Jonathan M. Smith. Edinburgh: Edinburgh University Press.

___. 1993. "The Lie That Blinds: Destabilizing the Text of Landscape." In *Place/Culture/Representation,* ed. James Duncan and David Ley. London and New York: Routledge.

Smith, Michael D., and Richard S. Krannich. 2000. "Culture Clash Revisited: Newcomer and Longer-Term Residents' Attitudes Toward Land Use, Development, and Environmental Issues in Rural Communities in the Rocky Mountains." *Rural Sociology* 65 (3): 396-421.

Tuan, Yi Fu. 1977. *Space and Place.* Minneapolis: University of Minnesota Press.

___. 1995. *Passing Strange and Wonderful: Aesthetics, Nature, and Culture.* New York: Kodansha International.

___1998. *Escapism.* Baltimore: The Johns Hopkins University Press.

Tolbert, Charles. et al. 2002. "Civic Community in Small-Town America: How Civic Welfare is Influenced by Local Capitalism and Civic Engagement." *Rural Sociology* 67 (1): 90-113.

Wallin, Luke. 1998. "The Stranger on the Green." In *Philosophy and Geography II: The Production of Public Space,* ed. Andrew Light and Jonathan M. Smith. New York: Rowman & Littlefield.

Wood, Robert, and John Handley. 2001. "Landscape Dynamics and the Management of Changes." *Landscape Research* 26 (1): 45-54.

Woolcock, Michael, and Deepa Narayan. 2000. "Social Capital: Implications for Development Theory, Research, and Policy." *World Bank Research Observer,* 15(2) 225-49.

Sociomaterial Communication, Community, and Ecosustainability in the Global Era

Richard Westra

Introduction

In an important intervention in the literature on environmental political economy, John Dryzek (1996) explains how debate over the possibility of constructing an ecosustainable future for human society has largely swirled around two perspectives on human nature. On the one side is *Homo economicus,* the instrumental, "rational" actor of neoclassical economics. And, on the other side, there is *Homo ecologicus,* the new, ascetic ecological subject of "Green" social theory. For Dryzek, it is abundantly clear that little space exists in instrumental rationality for the ecosensitivity required by a future environmentally responsible social order. On the other hand, he is concerned with the possible authoritarian political outcome of inculcating the new ecological subjectivity and the extent of the social regression entailed in certain Green visions of self-sufficient microcommunities. What troubles him to an even greater extent is the very endeavor of both sides to reduce the issue of environmental political economy to "psychology" in the first place. As Dryzek (1996: 30) posits, the need for social science derives precisely from the fact "that society, and social structure, are not reducible to psychology." Thus he proposes a twofold solution for thinking anew about the political economy of environmental sustainability. First, Dryzek (1996: 38) argues for "an ecological political economy [that] would concern itself with the structure, organization and operation of political–economic systems as they confront ecological problems." Second, Dryzek (1996: 28) argues for "human intersubjectivity and communication" as the nexus through which the processes of institution-building for the future eco-sustainable order are to be articulated. However, while I am in agreement with the above characterization of social science and the task of political economy, and accept that social "communication" broadly conceived has an important role to play in remaking our social communities, Dryzek's article never considers a set of deeper, more pressing questions that spring from his insights.

For example, what exactly is inferred with the vague term "political economic system" and what relation does it bear to the broader notion of forms of historical society such as *capitalism*? Does a touchstone exist for determining the range of possible political economic systems once such are defined—whether a particular system constitutes a progressive alternative to that currently in existence? Are there criteria we can adopt to assess the viability of proposed political economic systems to sustain human existence in the long term? How is the relationship between system viability and ecosustainability to be clearly specified? What sort of interface exists between various political economic systems and particular modes of social communication? Are specific types of society predisposed to certain communicative forms? In short, while the list could be extended, it is the view of this essay that as we engage in future directed thinking about improving the human condition and the ecosanctity of communities in which life is embedded, it is vital that answers be provided to questions like the foregoing. Of course, while an attempt to answer all the questions involved would be a project for a long book, it is the intention of this essay to make a contribution to the endeavor with the provision of a framework for making the key determinations over what I have dubbed *the material reproductive viability* of forms of human society and assessing the ecosustainable pedigree and progressive potential of these. The essay also introduces the novel concept of *sociomaterial communication* as a means of thinking creatively about institutional design of a future environmentally sound society. To accomplish its task, the essay will adopt the following procedure: First, the essay presents a unique perspective on political economy originated in Japan and demonstrates how through its study of the capitalist economy it yields all-important insights into material existence *per se*. Second, on the basis of the Japanese approach, the essay will examine the prospects for a materially viable ecosustainable future inherent in the trajectory of current globalization. Third, the essay will draw upon elements of Green theory and new ideas of socialism to offer rudiments of a model of a materially viable ecosustainable future.

Capitalism and the General Norms of Economic Life

No more pressing question exists today as we seek to move beyond the narrow confines of the debate over a possible Green subjectivity, toward understanding how "social structure" shapes social outcomes, than that of producing knowledge of the deep structure of the political economic system of capitalism. The work of economic historian Karl Polanyi is highly instructive in this regard in the way it differentiates between capitalism and precapitalist societies through the notion that the *economy* in capitalism tends to become "disembedded" from other realms of the social–the politics, religion, and so on– with which it had been intermeshed since the dawn of human society. It is the view of this essay, however, that Polanyi's conception is but an imprecise term for what Karl Marx had dubbed capitalist *reification*. What reification captures, according to Marxian theory, is not only the fact of the economy seeming to delink from the social, but that though capitalism is a socially and historically constituted social order, the economic in capitalism comes to take on a "life of its own" and *wields* the social, human beings, and the human life-world for its own self-aggrandizement–the augmentation of value. Yet it would remain the task of the Japanese political economist Kozo Uno (1980)[1] to draw out the ultimate implications of Marx's work for social science, the analysis of capitalism, and social scientific studies of material life in all human society.

It is no accident that economics emerges as a field of study only in the age of capitalism. For it is precisely the reification of economic life or the disembedding of the economy from the social that constitutes the ontological condition of possibility for economic theory. That is, as Polanyi, Marx, and Uno observe, precapitalist societies reproduce their economic life through varying forms of interpersonal relations of cooperation, dependence, or domination and subordination. Capitalism, however, dissolves such interpersonal relations of economic existence in its organizing of human material life in impersonal, society-wide integrated systems of self-regulating markets operating to augment value. In this fashion, and I will elaborate further upon this below, capitalism converts human relations of material life into "relations among things," and it is this conversion in the abstract working of self-regulating markets that "objectifies" them and renders economic relations "transparent" for theory to explore. Therefore, the social science implication of the study of capitalism is that political economy demands a specific cognitive sequence. That is, the unmasking of the abstract inner workings of *capital* not only offers the most fundamental critique of capitalism in its variant historical manifestations but also provides a window of opportunity for the elaboration of what Uno (1980: xix) dubbed "the general norms of economic life," norms that capitalism as well as all other human societies had to satisfy as conditions of their material reproductive viability. Focused upon from another direction, two important points flow from the foregoing. First, material economic life, something without which human society would be impossible, and the ontologically peculiar modalities of reproduction of that life by capital, are *not* the same thing. Second, for any economic theory of capitalism to prove its mettle, it must be able to demonstrate how capitalism is able to realize its chrematistic of augmenting value while simultaneously satisfying the general norms of economic life required by any form of human society.

When neoclassical economists refer to the market reaching a state of "general equilibrium," what they have in fact unwittingly hit upon is precisely the question of the economic viability of capitalism. I say "unwittingly" because as per Dryzek's work cited above, the neoclassical approach to economics, in terms of the suprahistoric instrumental rationality of individuals, suggests that there exists *no* distinction to be made between the chrematistic operation of the capitalist commodity economy and material life *per se*. Contrariwise, what the Japanese Uno approach to Marxism contributes to the development of economic theory and political economy is an elucidation of how under conditions of reification, where social relations of material life are converted into relations among things, capital is able to constitute a viable historical society. In the economic model Uno dubbed the *theory of a purely capitalist society* (TPCS)[2], there is agreement with neoclassical economics that market forces of supply and demand determine the relative prices of commodities. But what neoclassical economics does not recognize is that for this phenomenon to be meaningful in an economically substantive sense, it must be placed within the context of the historical conditions of specifically capitalist production. That is, all decision-making on the part of economic actors, particularly investment decisions by the capitalist in response to price signals, requires the commodification of labor-power to bear fruit. For it is only on the basis of the existence of a class of "free laborers" offering their labor-power on the market for capital to purchase, as but another input into the production process, that it is possible for capital to shift to the production of *any* good as per the changing pattern of social demand and opportunity for profit-making.

In other words, no human society could survive for long if it chronically overproduced or underproduced basic goods relative to the existing pattern of social demand. If, for example, there remained uncorrected a chronic misallocation of resources, principally human labor, from production of basic foodstuffs to production of iron and steel, the respective society would eventually collapse. And human history is littered with precisely such examples of societies the material-reproductive *modus operandi* of which could not ensure this fundamental requirement. The central operative principle of the capitalist market, then, is the *law of value*, which, under the constraints of capitalist social relations of production, works to prevent any chronic misallocation of social resources by ensuring that commodities embody only *socially necessary labor*. Socially necessary labor here refers to the peculiar means by which work in the capitalist commodity economy is validated. If capital deploys labor-power in the production of goods that are not in demand or in the production of goods with the operation of redundant technologies, the commodities will not be sold and profits not made and, from the perspective of capitalist society as a whole, such work will be deemed a waste. In this sense, the law of value "mediates" between the specifically capitalist commodity economic organization of material life and the production of use-values that constitute the basis of human material existence *per se*. And conceiving of a general equilibrium in the absence of this understanding of the material reproductive viability of capitalism is tantamount to the view that somehow that capitalist "market" could be decoupled from the capitalist mode of production.

Globalization and the Ecosustainability of Capitalism

Let us now move as promised to an unpacking of the earlier statement that the capitalist market constituted a domain of impersonal relations among things. For it is from here that spring the most profound insights into the ecosustainability of capitalism. When mainstream economics adverts to the economic "efficiency" of the capitalist market, it is highlighting the market's "costless" transmitting of economic information to economic actors in the way of market prices such that an "optimal" allocation of resources, captured in the notion of a general equilibrium, will be achieved. What the capitalist market is in fact doing here is performing a series of abstract "calculations" based upon value/price or *quantitative* criteria. Through the operation of the integrated system of self-regulating markets of capitalism, then, human beings in fact abdicate their responsibility for organizing their economic affairs to what amounts to an "extrahuman" force that reproduces their economic life only as a by-product of value augmentation. Recalling the discussion above, this chrematistic operation of capitalism is what the term "reification" is intended to capture. And, because value augmentation is an *abstract-quantitative* goal, we can say that at a most fundamental level capital, with its commodity economic material outcomes, is destined to conflict with *concrete-qualitative* human goals to the extent that the latter necessitate respect for the earth and the life-world within which long-term human existence must be embedded. Such is illustrated so vividly by the historical record of capitalism that displays how value augmentation can proceed extremely successfully through the production of noxious goods as well as those with the potential to destroy life itself. Paradoxically, therefore, while an unimpeded procedure of abstract market calculation is the basis of capital's ability to constitute a viable economic order, it is also the root of capital's profaning of the human life-world and the earth.

Of course, an objection to the foregoing might be that, whether portrayed in neoclassical terms or those of the Uno approach, the tending of the capitalist market toward the equilibrating of supply and demand is achieved, if at all, only with extreme difficulty and under a highly restrictive series of conditions in actual capitalist history. In addition, across the sweep of capitalist history economic outcomes are never as stark as depicted in theory; the harsh impact of abstract market forces is dampened by political means. Such intervention in the market is conceptualized by economists as the management of market *externalities*. Indeed,

I have argued (Westra 2003c) that it is precisely this managing of market externalities in capitalism that foregrounds the study of the capitalist state. And, through the prism of the level *of analysis* the Uno approach refers to as *stage theory*,[3] this essay will proceed to address the question of state *policy* in support of capital accumulation in terms of both the material reproductive viability of capitalism and its ecopedigree. The argument here in brief is that capitalism is marked by four world-historic stages of development—mercantilism, liberalism, imperialism and consumerism—each characterized by a dominant form of commodity production (wool, cotton, steel/heavy chemicals, and automobile/consumer durables, respectively), a geospatial core, and a stage-specific institutional architecture that constitutes the matrix through which capital accumulation proceeds and extra-economic supports for capital are articulated. Thus, the naming of stages of capitalism derives from the nexus of the core modalities of capital accumulation in each stage and the dominant state policies set out to support capital and manage the stage-specific externalities. From this perspective, if there ever exists a stage of capitalism that exemplifies the tendencies of capital captured in the TPCS, where the externalities the capitalist state was called upon to manage were minimal, it is the stage of liberalism characterized by capital accumulation of mid-nineteenth-century Britain. For example, characteristic liberal stage businesses were entrepreneurial and price competitive, technological advancements of *industrial* capital proceeded vis-à-vis reinvestment of business profits, labor-power was essentially unorganized, energy requirements were relatively meager, the pattern of social demand entailed the production of a cluster of quite simple use values (typified by cotton production), and so on.

On the other hand, the capitalist stage of consumerism typified by U.S. capital in the post-World War II period, and marked by the production of consumer durables (typified by the automobile), represents a huge departure from the market-equilibrating principles of capitalism captured in the TPCS or approximated by the *laissez-faire* industrial/entrepreneurial capitalism of nineteenth-century Britain. To manage the capitalist *mass*-production of such a relatively *complex* commodity cluster, consumerist capital could only at its peril leave accumulation to the vagaries of the market, and it increasingly adopted principles of economic programming and planning. For example, the massive investment costs of consumer durable production led to the further refinement of the corporate type of business structure that would strive to internalize transactions, become deeply involved in demand management, and cultivate a cooperative relationship with labor to keep interruptions of production to a minimum. The internationalization of production and finance characterizing consumerism not only intensified such trends but also contributed to corporate capital's need to coordinate each and every arm of business activity. As a noted business analyst put it: "Although no global corporation yet manages a planned economy on the scale of the former Soviet economy, they are coming closer"

(Korten 1995: 221). Paralleling the programming and planning activities of corporate capital was the rise of the consumerist state with a formidable policy arsenal at its disposal to support capital accumulation. The social wage, creation of effective demand (military, transportation infrastructure, and so on), and monetary, fiscal, labor, and trade policy are just a few of the well-known initiatives. The upshot of the foregoing is that, as with the augmentation of value, so the viable material reproduction of the economic community in the capitalist stage of consumerism becomes more and more dependent upon extramarket principles—a trend that, as we shall see below, set the course for a peculiar political economic outcome.

However, to refocus the argument back on the question of the ecoviability of capitalism, the reasons why the capitalist stage of consumerism emerges as the most environmentally unsound stage of capitalism hardly need any rehearsal. The mass consumption of consumer durables on which corporate accumulation depends necessitates that consumer demand for an ever-expanding and novel assortment of such goods remain virtually insatiable, thus generating rapid product obsolescence and mountains of waste. Further, the energy profile, mainly that of petroleum but also nuclear power coupled with the gargantuan energy requirements for electricity and transportation, has ravaged the biosphere and contains the specter of human annihilation. And global pollution has only been compounded by the increasing internationalisation of production, for both the latter's tendency to shift environmental problems around the world and the need it promotes to augment hyperpolluting transportation networks. All this of course saddles the consumerist state with the burden of managing not only the aforementioned burgeoning externalities of capital accumulation but also those now involving the very ecosustainability of the social community within which capital operates. Thus it is here we come to my final point in the discussion of the all-around viability of capitalism.

What passes for "globalization," the neoliberal inspired processes of national deregulation, financial liberalization, deindustrialization of capital's heartland, and so on, amounts to the fact that the consumerist state is increasingly abdicating its responsibilities for managing those externalities that capital had demanded of it to enable capitalism to continue as a viable economic order. That is, while capital depended upon the programming and planning of the capitalist state, as capital accumulation continued to slow across much of the capitalist heartland from the mid-1970s onward (Webber & Rigby 2002), capital also began to experience its infrastructure of extramarket supports as a constraint and sought to dismantle it and "free" itself from them. In opposition to neoliberal ideology that would have us believe the world economy is in transition to a new stage of market-equilibrating entrepreneurial capitalism, I have argued (Westra 2003c, 2004b) that given the persistence of the corporate/consumer durable production/high energy use/high waste material substructure of the world economy today, without the state performing its historic role as a *capitalist* state (one that manages the market externalities that necessarily encumber such an economy), we are instead witnessing a transition *away* from capitalism, a transition whereby other material reproductive principles, the contours and potential viability of which are not yet completely clear (though certainly entail authoritarian modes of labor and population control as well as forms of global apartheid), are emerging to fill the void. To this I would add these points, which portend bleak prospects for the ecoviability of the current political economy: the short-term and "off-ground" business outlook of the predominant economic actors (leading to the increasing dispersal of marketable productive capacity across the globe, where there is even less inclination on the part of capital to concern itself with the managing of any externalities); the fact that national and international regulatory systems for ecomonitoring have never kept in step with this trend of internationalization of production, and that those systems which were developed are now being progressively gutted; and the persisting emphasis upon consumption of throw-away consumer durables as the height of human fulfillment.

Sociomaterial Communication and Ecosustainable Communities

If, as has been argued, capitalism in both its fundamental incarnation and its historical development is not up to the task of providing human beings with a materially viable ecosustainable future, then what sort of sociomaterial arrangements can be marshalled for that purpose? First, it is strongly believed within this essay that current globalization offers no template for a progressive future. As elaborated upon elsewhere (Westra 2003c, 2004b), globalization is failing human society in terms of material reproductive viability. That is, with the world economic asymmetries in productivity, resource capacity, and so on, which together mitigate the rise of neoclassical "perfect" competition and the tending of the global economy toward a general equilibrium, there exists no coordinating economic principle to ensure that the variegated social demand for basic goods is met. Further, without the rise of a world capitalist state, no institution would be adequate to the task of managing the extensive externalities generated by this economy. The spiraling out of control of economic disparities, world hunger, disease, environmental distress, and so on, combined with the abject failures of effective global governance in all areas involved in the provisioning of "social goods," clearly portends the future.

What about the possibility of a form of socialist society providing the pathway to a progressive human future? It must be accepted at the outset that the Soviet-style model of socialist development was a disappointment with regard to the cumbersome Soviet- style centralized state planning apparatus, which only barely satisfied social demand for basic goods (and did so with a chronic waste of social resources, the test of material reproductive viability), and its dismal environmental record is beyond dispute. Without entering into esoteric debates over whether in fact the Soviet-style exemplar was actually *socialism,* let us examine recent work in the building of socialism, theoretically. To begin with the question of material eproductive viability, models have been developed that clearly demonstrate how computers may be utilized to perform market calculations and obtain equilibrium resource allocations combined with redistributive social outcomes(see Cockshott & Cottrell 1997, 2003). Paralleling this work are models confirming that reliable data need not emanate from a central plan but instead derive from an iterative channeling procedure through decentralized participatory, democratic micro, meso, and macro decision-making bodies (see Albert & Hahnel 1991). While such work suggests an answer to the question of material reproductive viability, and also promises to purge socialism of its authoritarian propensities, the issue still remains of the ecopedigree of such a mode of economic planning. Far from promoting overly ambitious utopian plans for social change, socialist approaches to the future have remained a prisoner of capitalist thinking about the realm of possibility for organizing human economic affairs (Westra 2002a, 2002b, 2004a). That is, whether in its discredited centralized authoritarian example or in the more palatable democratic decentralized version, conceptions of planning have been framed in terms of essentially substituting or simulating the operation of the society-wide self-regulating market of capitalism. However, the very characterization of the primary economic problem of socialist society as one of economic *calculation* reduces the possibility of socialism to an abstract technical issue; notwithstanding the democratic infusions, it is questionable whether *society-wide* economic coordination based solely upon the planning principle will ultimately engender the sort of sensitivity to human need, the human life-world, and the earth that socialists are genuinely seeking. There is also the related question of realizing *sociomaterial betterment,* which demands (among other things) that new forms of socialist *motivation* be fostered to strip work of the effects of alienation,[4] though such also affect the shoring up of ecosensitivity.

If we shift the focus away from simulating capitalism, and toward the creative study of the varying means through which the general norms of economic existence can be satisfied, as the Uno approach indicates, building the future society both theoretically and in practice will involve an entirely different set of questions. For example, from the TPCS, which captures capital in its most fundamental incarnation as a reified commodity economy, flow the most vital insights into what must be *undone* in our economic existence to purge it of capital's crippling effects and remake our economic lives through personal relations among human beings rather than capitalist relations among things. Similarly, the role of stage theory, particularly with regard to consumerism, is not intended to confirm what capital has bequeathed to a socialist future but to analyze the stage structures of accumulation and noneconomic supports so as to be clear on precisely what is necessary to dismantle such and the economic distortions they entail and replace them with new forms of socialist economy. This, however, brings us to the signal question of the institutional vehicle to realize socialist aims of material reproductive viability with equity and ecosustainable community life.

Green social theory, animated by work such as that of Schumacher (1999), has long demanded the devolution of economic life to self-sustaining microcommunities. Besides the points made by Dryzek in the introduction, critics (see Martell 1994; Pepper 1993) argue that ecodegradation exceeds the capacity of even nation-states to remedy, thus making it difficult to comprehend how current problems could be dealt with solely in an autarchic context. As well, it is pointed out that there in fact is no inherent necessity for microcommunities to automatically generate sound environmental outcomes; not only Schumacher but the genre of Green writing in this area has been vague on how communities might coordinate ecopolicy to ensure that the results of potentially unsound activities are not blithely passed along to others. This problem brings us back to the thorny issue of how new ecosubjectivity might be cultivated. Further, given the vast resource asymmetries that exist worldwide, it is not even clear that Schumacher's autarchic communities would be material reproductively viable.

My own work has also made the claim that a genuine socialism must rethink the question of economic *scale* (Westra 2002b, 2004a). But the consideration of scale put forward involves placing the debate within the context of the political economic study of material reproductive viability: Satisfying the general norms of economic life, the question of sociomaterial betterment, and the broader problem of economic coordination are elided in Green theory. In the socialism I have advocated, then, for the all-important satisfying of social demand for basic goods, socialist strategy must promote the creative redesign of communities in a form that rebuilds the connection between productive activities and community consumption that centuries of capitalism have sundered.[5] For it is the commodity-economic cultivated disinterest of workers to *what* is produced and their indifference as consumers in capitalist society to *how* goods are produced that nurture environmental neglect. And, as noted above, it is difficult to envision how, if socialist societies co-opt the capitalist division of labor in a *society-wide* plan, that even factoring in the effects of decentralized decision-making, ecoviability would be significantly improved.

The question of scale, however, is inexorably tied to the problematic of economic coordination. That is, the framing of such, solely in terms of economic calculation, must be abandoned for the reason stated above: that it imparts to socialism the one-sided, technical approach to the transmitting of economic information characteristic of the capitalist commodity economy. What I have maintained is that the question must be restated in terms of the broader category of *sociomaterial communication,* of which price calculations of markets or equilibrium-achieving iterative plans are only a few of many possible forms. Sociomaterial communication or economic coordination is something not addressed in Green theory. In the community model I have proposed, economic intercourse may involve (small-M) markets–markets that, as Polanyi and others point out, existed at the interstices of ancient society, trading in luxury goods, peripheral to the central principles of interpersonal economic relations marking those societies – that could operate according to either principles of need exchange or local exchange and trading systems (LET) with local currencies. Community life might also reinstate forms of barter or reciprocity in economic relations. These forms of sociomaterial communication would coexist with a variety of planning mechanisms and ownership forms, though the "plan" must be to efface *quantitative* considerations in the production of as many categories of goods as possible and replace such with *qualitative* use-value considerations. For it is, I believe, the quantitative considerations in society-wide equilibrium planning models that contain environmentally unsound residues of capitalist production modalities. And placing qualitative considerations at the heart of economic life also entails addressing important issues such as production and product choices, work alienation, appropriate technologies, and so on.

To offer genuine sociomaterial betterment of the sort that would animate mass public interest in change, the future order must not only seek to maintain elements of cosmopolitan society that have marked the modern era: There are certainly categories of goods, including environmentally sound forms of mass transportation and even modes of environmentally friendly energy delivery, the production, infrastructure, and operation of which necessarily demand economies of scale and regularizing and coordinating of economic activity that will far outstrip the capacities of community production. In the brief sketch that follows, I will build on earlier work (Sekine 1990, 2004; Westra 2002b, 2004a) to offer an institutional remedy entailing the development of two basic configurations of economic community; each is institutionally geared to managing production of specific types of goods. Quite simply, "qualitative goods" (those that lend themselves to qualitative production concerns, including agricultural production, and depending on access to raw materials, various types of apparel, household sundries, furniture, building materials, and so on) could be produced in the sort of communities described by Schumacher. The production of "quantitative goods," on the other hand, including consumer products that lend themselves to standardized production methods, "heavy" goods (both producer and consumer), infrastructure materials, and so on, could be managed by a "state" sector composed of larger production organizations. Of course, the distinction between these categories is not firm. Responsibility for labor-intensive final assembly of numerous products from the quantitative side of the ledger can be allocated to individual communities. Its force is rather to offer a means of disrupting global "commodity chains" shaped by the demands of value augmentation and to open the door to creative thinking about how varying principles of economy may be articulated and matched with appropriate forms of sociomaterial communication and ownership.

The economy of the qualitative goods-producing community would operate with a combination of principles of economy and communicative forms noted above. Congeries of such communities, bound by geographic proximity, potential economic connections, political interest, and so on, would link up with the state organizations producing quantitative goods that operate according to principles of participatory planning. The state sector will have its own currency, exchangeable with the local currencies of the communities with which it interfaces, as well as with the currencies of other "states," though again, the issue of whether the construction of a socialist commonwealth of states would utilize the configuring of current states as a template is something that will have to be dealt with in practice. Finally, cities linked with the local communities and state sector may, for the foreseeable future, remain the sites of administration and governance and would also use the state currency as means of exchange. The ecosustainability of each form of community, as well as the patterns of interconnections, could be assessed according to techniques of the "*ecological footprint*" (Rees & Wackernagel, 1995; Chambers, Simmons & Wackernagel, 2000) where the environmental "carrying capacity" of the region is regularly measured and deficits are democratically addressed. The choices among potential nonprivate ownership forms for the state sector are quite varied. What might optimally be adopted is a system that maintains the form of the modern corporation and through which individuals and enterprises of local communities, state sector workers, professionals, and technicians in the urban areas are all established as shareholders, where shareholding involves constraints placed upon intergenerational transfer, accumulation and sale, and so on. Of course, this model assumes that qualitative goods communities as well as the state sector significantly rethink and ultimately alter the use-value and energy profile from that currently in existence.

To be sure, the precise delineation of the future society cannot be made without considering a plethora of issues including geospatial locale, level of techno-economic development, regional natural resource endowment, relative size of urban centers, the sorts of power and wealth asymmetries to be redressed, and so forth. Nevertheless, in drawing together the threads of the argument for a material economically viable and ecosustainable socialism, the signal points are as follows. First, the greatest possible range of basic goods must be produced through democratic negotiation at the common ownership community level, the specified size of which is to be determined in each concrete transformatory situation. Second, the mode of sociomaterial communication will largely involve forms of reciprocity and face-to-face need exchanges of goods. Markets of course have a place here but are marked by use of local currencies and play a benign role in distributing community products as opposed to being the loci of unrestrained competitive price-takers operating on the basis of private property. Third, producer and heavy consumer goods with economy of scale and increased socialization of investment requirements will be the production responsibility of a state sector deploying democratic iterative planning as its mode of sociomaterial communication. This state sector, owned by the communities it interfaces with and utilizing a single central currency, is also at the outset bifurcated between an administrative urban core and production regions. In short, the tricommunity structure materializes a circular-flow economic model to meet the criterion of economic viability. Its ecosustainable pedigree derives from the technique of the ecological footprint referred to above. The marshalling of varying modes of sociomaterial communication guarantees the sensitivity of the social order to the widest possible range of economic needs. Ultimately the goal of such arrangements is to break down the impacts of commodity economic-induced asymmetries between town and country, manual labor and administration, industry and agriculture, and so forth.

In the light of the foregoing, finally, a brief comment on human "rationality" is in order. It should be evident that within the context of local communities where human beings are "producing for themselves," self-motivation, with all the implications such carries for ecosustainability and sociomaterial betterment, will be attained with little difficulty. On the other hand, work alienation, the domain of instrumental rationality, can only be completely overcome in the quantitative goods or state sector under conditions where production is fully automated. One way of approaching this difficulty is to create a system for the democratic rotation of labor forces and working families from communities through state production facilities and cities. This would cultivate the multidimensionality of human beings, a component of genuine socialism's promise of socio-material betterment. And it would serve to promote ecosustainability in that people would have an incentive to maintain the lived environment and eco-sanctity of each of the communities in which they reside. Such a system could then be extended within a wider socialist commonwealth that facilitated increasingly open emigration and immigration in efforts to overcome the apartheids, political, economic, and environmental, that mark the current global era.

Conclusion

The issue of social change is complex, and while it is not clear what configuration of social constituencies will rise to undertake the task, the time has arrived when it can no longer be avoided. To be sure, it will certainly take at least a few generations to efface the residues of the capitalist commodity economy and reverse much of the damage that this economy has hitherto inflicted upon the life-world and earth. Some of the work, however, can begin immediately at the level of community life, as the global proliferation of LET and community-based environmental movements worldwide have displayed. The plan must be to disentangle our economic existence from capitalism and many of its deleterious practices and construct the new society. Whether we call it socialism or give it another name, it must be built on the basis of knowledge that the political economy research agenda of the sort introduced here offers on the study of the general norms of economic life and material reproductive viability, sociomaterial communication, and ecosustainability. Only in this way will our endeavors hold out the promise of overcoming the disappointments of the projects for change of the past.

Notes

1. Uno (1980) is an English translation of an abridged version of a two-volume work, both available only in Japanese. Signal monographs in English include Sekine (1997), which constitutes a two-volume recasting and refinement of Uno's writing (Sekine was a student of Uno), and Albritton (1991). The article literature is voluminous, but for a succinct summary of the approach written from a comparative perspective, see Westra, (1999).
2. Without engaging in the esoteric debates that plague the field of political economy, the TPCS derives from Marx's writing in his three-volume *Capital* but arguably completes that project which Marx left unfinished and does so in the modern language of economics and includes debate with neoclassical adversaries.
3. Quite simply, the Japanese Uno approach argues against utilizing abstract theory directly to "model" historical outcomes. It argues, rather, that political economic study of capitalism requires three levels of analysis where the movement in thought from the abstract TPCS to what Unoists refer to as *historical-empirical analysis* is "mediated" by a stage theory of capitalist development. For recent work on stage theory see Albritton (1991) and Westra 2003a, 2003b).
4. Quite simply, the concept of sociomaterial betterment is intended to capture the shifts in structures of compulsion for work, economic empowerment, and quality of material reproductive life in all its multidimensionality and so on from that existing in capitalist and precapitalist societies. See Westra, (2002a)
5. In this sense, I wholeheartedly agree with those who argue htat the division of labor in capitalist society is shaped by *capitalist* social relations of production and is not simply a "technical" development of "industrial" society. See, for example, Gough and Eisenschitz (1997).

References

Albert, Michael, and Robin Hahnel. 1991. *Looking Forward: Participatory Economics for the Twentieth Century.* Boston: South End Press.

Albritton, Robert. 1991. *A Japanese Approach to Stages of Capitalist Development.* Basingstoke: Palgrave Macmillan.

Chambers, Nicky, Craig Simmons, and Mathis Wackernagel. 2000. *Sharing Nature's Interest: Ecological Footprints as an Indicator of Sustainability.* London: Earthscan.

Cockshott, W. Paul, and Allin F. Cottrell. 1997. "Value, Markets and Socialism." *Science & Society* 61 (3): 330-57.

Cockshott, W. Paul, and Allin F. Cottrell. 2003. "Economic Planning, Computers and Labor Values." Paper presented at the International Conference on the Work of Karl Marx and Challenges for the XXI Century, Havana, Cuba, May 5-8.

Dryzek, John S. 1996. "Foundations for Environmental Political Economy: The Search for *Homo Ecologicus?*" *New Political Economy* 1 (1): 27-40.

Gough, Jamie, and Aram Eisenschitz. 1997. "The Division of Labour, Capitalism and Socialism: An Alternative to Sayer." *International Journal of Urban and Regional Research* 21 (1): 23-37.

Korten, David C. 1995. *When Corporations Rule the World.* West Hartford, CT: Kumarian Press.

Martell, Luke. 1994. *Ecology and Society: An Introduction.* Amherst: University of Massachusetts Press.

Pepper, David. 1993. *Eco-Socialism.* London: Routledge.

Rees, William E., and Mathis Wackernagel. 1995. *Our Ecological Footprint: Reducing Human Impact on Earth.* Gabriola Island, BC: New Society Publishers.

Schumacher, E. F. 1999 [1973]. *Small Is Beautiful.* Vancouver, BC: Hartley and Marks.

Sekine, Thomas T. 1990. "Socialism as a Living Idea." In *Socialist Dilemmas, East and West,* ed. Henryk Flakierski and Thomas T. Sekine. New York: M. E. Sharp.

___. 1997. *An Outline of the Dialectic of Capital.* London: Palgrave Macmillan.

___. 2004. "Socialism Beyond Market and Productivism." In *New Socialisms: Futures Beyond Globalization,* ed. Robert Albritton, John R. Bell, Shannon Bell, and Richard Westra. London: Routledge.

Uno, Kozo. 1980. *Principles of Political Economy.* Atlantic Highlands, NJ: Humanities Press.

Webber, Michael J., and David L. Rigby. 2002, "Growth and Change in the World Economy Since 1950." In *Phases of Capitalist Development: Booms, Crises and Globalizations,* ed. Robert Albritton, Makoto Itoh, Richard Westra, and Alan Zuege. Basingstoke: Palgrave Macmillan.

Westra, Richard. 1999. "A Japanese Contribution to the Critique of Rational Choice Marxism." *Social Theory and Practice* 25 (3): 439-69.

___. 2002a. "Phases of Capitalism and Post-Capitalist Social Change." In *Phases of Capitalist Development: Booms, Crises and Globalizations,* ed. Robert Albritton, Makoto Itoh, Richard Westra, and Alan Zuege. Basingstoke: Palgrave Macmillan.

___. 2002b. "Marxian Economic Theory and an Ontology of Socialism: A Japanese Intervention." *Capital & Class* 78: 61-85.

___. 2003a. "Phases of Capitalism, Globalizations and the Japanese Economic Crisis." In *Turbulence and New Directions in Global Political Economy,* ed. James Busumtwi-Sam and Laurent Dobuzinskis. Basingstoke: Palgrave Macmillan.

___. 2003b. "Social Theory, Economic Crises, and the Japanese Political Economy: A Review Article." *Review of International Political Economy* 10 (2): 363-73.

___. 2003c. "Globalization: The Retreat of Capital to the 'Interstices' of the World?" In *Value and the World Economy Today: Production, Finance, and Globalization,* ed. Richard Westra and Alan Zuege. Basingstoke: Palgrave Macmillan.

___. 2004a. "The 'Impasse' Debate and Socialist Development." In *New Socialisms: Futures Beyond Globalization,* eds. Robert Albritton, John R. Bell, Shannon Bell, and Richard Westra. London: Routledge.

___. 2004b. "Globalization and the Pathway to Sociomaterial Betterment." *Review of Radical Political Economics* 36 (3) 381-90.

The Prospect of Sustainability in the Culture of Capitalism, Global Culture, and Globalization
A Diachronic Perspective

Snježana Čolić

Introduction

Recent analyses of capitalism have focused on the cultural practices of consumption. While all human beings consume, a central preoccupation of advanced capitalism is consuming. Moreover, consumption has become the cultural *telos* of capitalism. As Tomlinson (1992: 122) suggests, what has cultural significance are the high levels of consumption in advanced capitalist societies. For these societies introduce a particular set of meanings people attach to their consumption practices and to the significance of such practices for their sense of purpose, happiness, and identity.[1] Of course, such practices are not static but change over time. What has often been overlooked in cultural studies of consumerism is the *dynamic* nature of culture. Consumption is not a universal process; rather, as with other forms of global knowledge conveyed through the culture industries, consumption practices are situated activities. As an encounter between the materiality of cultural commodities and the cultural formation of a consumer, consumption occurs in a particular context, namely a locality. Global forces like mass consumption display their effects in particular locales; therefore, local realities can no longer be thought outside the global sphere of influence.

Similar to consumption, globalization is not simply a process of exporting "sameness," as Storey (2003) suggests. Some critics argue that globalization can be understood, as Pieterse (1995: 45) points out, "as a process of hybridization which gives rise to a global melange," while others regard it as a process of homogenization. The latter view, however, overlooks the countercurrents, namely the impact non-Western societies are making on the West and on one another. Storey (2003) asserts that globalization produces two contradictory effects, *sameness* and *difference;* there is a sense that the world is becoming similar as it shrinks under the pressure of time-space compression, but also that it is characterized by an increasing awareness of difference. Globalization has thus made the notion of a neatly bounded sociocultural isolate even more untenable. Therefore, anthropologists cannot focus on a spatial unit merely as a self-contained isolate. Colonial and capitalist interventions are part of the picture, as are earlier migrations and histories. This essay will explore how global culture has come to eclipse local knowledge, especially with respect to resource needs, and has thereby moved localities to embrace more universal consumption practices.

Diachronic Perspective

Our lives and personal identities are not a static set of circumstances but rather a process. However, contemporary political discourse will often frame national culture and identity as being "frozen in time," thereby concealing complex historical processes. Sorting out the definitive features of "our culture," therefore, becomes highly problematic, since, as Tomlinson (1992) suggests, the contents of "our culture" likewise continually shift over time. As a consequence, "our culture" is never purely "local produce," in the metaphorical sense, but contains traces of previous cultural borrowings that then become "naturalized." Similarly, language is full of imported idioms that eventually become naturalized.[2] In other words, what counts as "local," and therefore "authentic," is not a fixed content but is subject to change and modification as a result of the "domestication" of imported cultural goods (Ang 1996).

Following from this, the discourse of cultural imperialism can be viewed in terms of different configurations of global power, from imperialism, which characterized the modern period up to the 1960s, to contemporary globalization, which as Tomlinson (1992) suggests is a far less coherent or culturally directed process. Despite its political and economic ambiguities, the idea of imperialism contains the notion of the *intended* spread of a social system from one center of power across the globe. By contrast, the idea of globalization suggests *interconnection* and interdependency of all global areas, which occurs in a far less purposeful way, through economic and cultural practices that do not, of themselves, aim at global integration but nonetheless produce it.

Globalization implies complex relations that characterize the contemporary world. Underlying these relationships is what Harvey (1989: 240) calls "time-space compression," or a sense that the world appears to be shrinking under the impact of the new media, such as satellite television and the Internet, that facilitate the extension of social relations across time and space. So, time and space no longer dictate the range of our relationships. For Storey (2003), being near or being distant no longer organizes with whom we communicate. When viewed as a process of global compression, as Robertson (1990) suggests, the world appears to be united by global culture to the extent that it is regarded as a single place. Framed in this way, globalization tends to weaken the cultural coherence of whole nations, including the economically powerful imperialist states of previous eras (see Tomlinson 1992: 175). Globalization also undermines the sense of coherence and unity of particular societies and their cultures. Global developments that promote cultural dislocations, as Urry (1989) notes, include the growth of multinational corporations, whose annual earnings dwarf the national income of many nations, and the growth of the media of mass communication, but also the possibility of technological disasters that know no national borders.

The cultural experience of people caught up in global processes is likely to be one of confusion and uncertainty, along with the perception of powerlessness. When people find their lives to be increasingly controlled by forces beyond the influence of those national institutions that form the perception of their specific polity, their accompanying sense of belonging to a secure culture is eroded. There is simply no way in which the legitimacy of those immensely powerful global economic forces can be established within the existing political framework of nation-states. We cannot vote *in* or *out* multinational corporations or the international market system, and yet, as Tomlinson (1992) suggests, these seem to have more influence on our lives than the national governments we do elect.

Accounts of the cultural experience of globalization, similar to those describing "postmodernity," point to a general sense of cultural insecurity. The prevailing mood of postmodernity is uncertainty, paradox, the loss of moral legitimacy, and cultural indirectness. For Jameson (1984), postmodernity is the cultural logic of the third great phase of capitalism–late capitalism–beginning after World War II. After the earlier expansions of, first, the national markets and then of the old imperialist system, the latest expansion of capitalism produces a truly global system that can be seen not only in the complex networks of international finance and multinational capitalist production but also in the spatial context of the cultural experience it produces. People's experiences are shaped by processes that operate on the local and global levels; however, very few fully understand how global processes shape their lives. The reality of the networks of global technology that influence our lives, such as computers shifting capital around the globe in seconds, can be only dimly grasped in cultural terms. This is because none of us actually dwell in the global space where these processes occur; for Tomlinson (1992), an information technology network is not really a *human* space. Our everyday experience, then, is necessarily local, yet it is increasingly shaped by global processes.[4]

The cultural space of the global is one to which we are constantly referred, particularly by the mass media, but also one in which it is very difficult to locate our own personal experience. For today's global culture is tied to no place or period. It is contextless and, according to Smith (1990: 177), "a true melange of disparate components drawn from everywhere and nowhere, borne upon the modern chariots of global telecommunications systems." Widely diffused in space, global culture is cut off from any sense of the "past"; therefore, it has no historical identity. It is here, now, and everywhere, and for its purposes, the past serves only to offer some decontextualized example or element for its cosmopolitan patchwork. The preeminently technical nature of its discourse conveys this sense of timelessness. Eclectic, universal, timeless and technical, global culture is both viewed and felt as a constructed culture (see Smith 1990).

While nations can be understood as having historic identities, or at least deriving closely from them, global culture fails to relate to any such historic identity. Unlike national cultures, global culture is a culture without memories. There are no "world memories" that can be used to unite humanity; the most global experiences to date–colonialism and the two world wars–can only serve to remind us of our historical cleavages. Smith (1990) argues that should nationalists have suffered "selective amnesia" in order to construct their nations, the creators of a global culture would have to suffer from "total amnesia" to have any chance of success. Therefore, a central difficulty in any attempt to construct a global identity and hence a global culture is that collective identity is always historically specific; it is based on shared memories and a sense of continuity across generations. To believe that the techno-economic sphere will provide the conditions and therefore the impetus and content of a global culture is to be misled once again by the same economic determinism that dogged the debate about "industrial convergence" and to overlook the vital role of *common historical* experiences and memories in shaping identity and culture. We are still very far from mapping out the kind of global culture and cosmopolitan ideal that can truly supersede a world of nations, each cultivating its distinctive historical character and rediscovering its national myths, memories, and symbols. A world of competing societies, seeking to improve their comparative status rankings and enlarge their cultural resources, as Smith (1990) suggests, affords little basis for global projects, despite the technical and linguistic infrastructural possibilities.

Critiques of cultural imperialism can be regarded as appeals for a more reflective or inner dimension of cultural experience in a globalized system. In this respect, many critics link globalization processes with cultural demands for localization. Jacques (1989: 133) sees "a new search for identity and difference in the face of impersonal global forces, which is leading to the emergence of new national and ethnic demands." As these demands seem to be occurring everywhere in the world, a structural reorganization of the way in which cultural goals become defined and enacted will be required. This implies a deconstruction of the "autonomized" global institutions of late modernity. However "autonomized" and abstractly powerful those institutions seem to be, as Tomlinson (1992: 178) cautions, "we cannot allow ourselves to think of them as unchangeable." Even in the face of globalization, Milton (1997: 491) views all cultures as providing "knowledge, values, assumptions, goals and rationales which guide human activity. This activity yields experiences and perceptions, which shape people's understanding of the world. The process is not unidirectional, but dialectical." This line of thinking follows from that of Vico and Dilthey, who posit that the province of the historical is, in the end, the province of human activity. Ultimately, the shape of the human world, under the pressures of globalization, can only be fully comprehended as a function of cultural will.

Prospects for Sustainability

Critics of global capitalist culture have asserted its essential unsustainability. Bodley (2001: 71) views global capitalism as an unsustainable "culture of consumption." Schmookler (1991: 19) sees feelings of "unfulfilled longing" underlying the culture of mass consumption, as "advertising promises that the goods we can buy carry with them the states of consciousness we desire and in which the broken promise of each purchase leads to new yearnings. So, people become addicts, willing participants in an economy based on the premise of growth without limit." Wallerstein (1990: 54) posits that the greatest contradiction of the world's capitalist system is its failure to admit that the system, as it is constructed, cannot expand forever, stating that for "every theory of limitless expansion is a hazardous paradise, and impossible in the real world." To explain the latter claim, it is worth recalling some facts. Economic growth is the very essence of global capitalism, and, as a rule, it follows that there will be increased consumption. Also, perpetual consumption is very dubious, as is the culture of consumption itself, which, as Sklair (1991: 41) suggests, "proclaims that the meaning of life is to be found in the things that we possess ... [and] ... to consume, therefore, is to be fully alive, and to remain fully alive we must continuously consume."

Consumerism is a key element in the ideology of the global system. Bodley (2001) asserts that the scale and organization of commercial exchanges that support consumption practices serve to concentrate social power and thereby reduce sustainability. Daly and Cobb (1989) view the weakness of perpetual-growth economic models that underlie consumerism, notably in their attempt to externalize, and thus neglect, the physical and biological realities upon which human communities depend. So, as Schmookler (1991) points out, the market shapes not only our image of human nature but also the human reality within its grasp, while the materialistic appetite of the West serves as the engine of our environmental destructiveness. Sklair (1991) views the global capitalist system as driven by a powerful, commercially generated ideology that convinces people that perpetual economic growth will benefit everyone; however, developments thus far indicate that this will not happen. Global capitalism is aimed at short-term profit for some rather than long-term social benefit for many, according to Fox (2001). Wallerstein (1990) calls this a "myth of the rising standard of living" because it masks the realities of poverty, environmental deterioration, and the unequal economic relationships between rich and poor countries.[6]

The intrusion of consumerist-driven commercial societies into the economies of poorer nations is particularly disruptive, as these wealthy societies not only promote individualism and wealth inequality but, as Bodley (1999: 7) states, "also convert resources into commodities." The consequences of such growth and development are more than evident. Many nonindustrialized countries had their territories, resources, and autonomy expropriated because of the demand for resources emanating from industrial nations. For Bodley (2001), resource depletion raises many troubling questions about the prospect of sustainability within the neoliberal model of capitalism and the culture of consumption. The Brundtland Commission's (WCED 1987) appeal to redirect global economic goals toward sustainability is therefore understandable. On the other hand, Pearce and Warford (1993) point to studies claiming that simply correcting "economic distortions" would make growth sustainable; this view is dubious because, as Bodley (2001: 87) states, "while development may indeed become sustainable, growth, by definition, cannot be sustained indefinitely."

The concept of sustainable development raises a number of critical questions, not only about the relationship between growth, progress, and sustainability but also about the very idea of development. For Gadotti (2003: 3), development has a "very precise context within a progressive ideology rooted in notions of history, economics, society and the human being himself." The term has been used within a colonizing vision, according to which countries were designated as being in one of three "worlds," namely "developed," "developing," and "underdeveloped." The criteria for development were based on standards of industrialization and consumption in each country. In this way, the goals of development have been imposed by the neocolonialist economic policies of the so-called developed countries, resulting, in many cases, in a vast increase in misery, unemployment, and violence. Therefore, it is not surprising, as Gadotti (2003) suggests, that many people have reservations concerning the notion of sustainable development.

Likewise, the idea of progress has increasingly come to be analogous with that of economic progress to the detriment of other aspects of sociocultural life. Progress has thus been reduced to quantitative growth, and the quantifiable indicators of growth have become indicators of progress: "one type or one indicator of progress has been turned into a natural notion or natural state," as stated by Kalanj (1994). It follows from this that development must be conceived in a more anthropological and less economic-centric form, as Čolić (2002), Gadotti (2003), Eder (1990), and Kalanj (1994) have argued. According to its anthropological definition, "development" is "progress" and not regress; it is production, creation of something new, and not destruction. A formally logical conclusion would be that contemporary development, insofar as it contains elements of unsustainability, as Lay (1992) notes, is not development at all.

Following this logic, the notion of progress should be freed from the idea of subjugating nature and, as Kalanj (1994: 77) suggests, should be "thought of both in terms of material and spiritual, quantitative and qualitative, facts and values. Distortion happens when only one side of that complex notion is taken into account, which then aims to encompass, regulate and legitimize the whole notion." If we are to liberate this notion from the Western model of progress, based as it is on the subjugation of nature, we should, as Eder (1990: 79) says, "generalize moral action across nature and culture." This would entail undergoing a change in our currently held conceptions of nature, and then employing a form of morally grounded practical reasoning in our relation to nature. So the real problem is how best to conceptualize genuinely sustainable development using this form of practical reason.

The acceptance of sustainable development goes hand in hand with the rejection of orthodox economic models that imply insatiable individual human desire for commodities as the primary human motivation, as Bodley (2001) suggests. With this comes the sustainable notion that "relative wealth" within a society, as Daly and Cobb (1989: 87) state, "may be more important than absolute levels of economic wealth." Bodley (2001) claims that any culture that continues to accelerate its patterns of consumption in a fixed environment will eventually be forced into making trade-offs between environmental quality and continuation of its consumption pattern. For him, the most critical anthropological question is: What cultural forces drive the present consumption patterns of the consumption culture? It should be noted here that overconsumption is not an innate human trait; it is culturally determined. High rates of consumption, or lack of cultural limits on consumption, relate to social stratification. Perpetual wealth accumulation and increasing consumption are intrinsic to a global culture that concentrates the greatest power in commercial organizations. Hence, this world system disproportionately serves the personal interests of the global elite who control the great multinational corporations and financial institutions, as many authors (Bodley 2001; Fox 2001; Gray 2002; Wallerstein 1986) have suggested.

When discussing sustainable development, one should note that the concept does not speak against economic growth, *a priori*. Rather, it advocates a change in the *quality* of growth, thereby promoting further development of the quality of life, meeting basic existential needs but also respecting the limitations imposed by the laws governing the living world. Definitions of sustainable development often contain the claim that current generations should not create such conditions that would make the needs of the future generations difficult or impossible to meet. Sustainable development is therefore a normative concept because, as Cifrić (2001: 165) states, it "usually says that it is possible to go on, and speaks about how to go on." It thus requires an ethical subject, as Küng (1997), Cifrić (2001), and Gadotti (2003) each suggest, one who changes the reality and, at the same time, acts responsibly. A culture of sustainability also presumes a pedagogy, in Gadotti's (2003) estimation. Viewed from a critical perspective, sustainable development requires an educational component, as the preservation of the environment depends upon an ecological conscience, and the formation of this conscience depends upon education.

In principle, sustainability and the culture of capitalism are not necessarily incompatible. However, these two concepts appear to be irreconcilable within the context of capitalistic globalization, as Gadotti (2003) asserts. Furthermore, in current debates, globalization as an historical process, ongoing for centuries, is being conflated with the political project of the free market and its neoliberal ideology worldwide. The dominant neoliberal ideology of global *laissez-faire* based on the market and on uncontrollable consumption, for Gray (2002), is in collision with permanent and vital human needs, particularly those of security and social identity. We will therefore need to distinguish between the two kinds of globalization, as in Gadotti's (2003) dichotomy of "competitive globalization" and "solidarity globalization." While terminologically similar, the two convey opposite logics. The first type subjugates states and nations and is governed by capitalist interests. Within competitive globalization, market interests are placed above human interests, and the interests of nations are subordinated to the corporate interests of large transnational firms. By contrast, the second type is carried out through a reorganization of civil society that would presumably bring about a "globalization of citizenship."

The prospect of a "global free market" does not represent a model of natural development rooted in society. On the contrary, it hinders legitimate forms of development suited to the traditions and needs of particular societies, fostering within them increasing resistance. Global competition favors the forms of capitalism that are socially least sensitive and result in damaging consequences for the stability and cohesion of societies, which, for Gray (2002), create fertile ground for the destruction of mechanisms of solidarity. For this reason, a key question is that of coexistence between the market and solidarity.5 Currently, even in a "free market," the instruments of the global market have broken free of social control. A possible path out of the present situation would be heightened awareness and stricter control by citizens over the state and the market, which, for both Lay (1992) and Gadotti (2003), would lead to a stronger civil society and the ability to control development. For them, sustainable development makes sense only within an economy of solidarity, that is, one not driven merely by profit; such an economy would presumably give rise to "solidarity globalization." Within this social universe, "global citizenry" could conceivably realize a sense of "coexistence in diversity."

Conclusion

Through a diachronic analysis, I have argued that we can no longer consider societies to be isolated, self-maintained systems, or cultures to be integrated totalities, in which each part contributes to the survival of an organized autonomous and constant whole. My own perspective follows from that of anthropologist Eric Wolf (1982: 390-91), who states that:

> There are only cultural sets of practices and ideas, put into play by determinate human actors under determinate circumstances. In the course of action, these cultural sets are forever assembled, dismantled, and reassembled, conveying in variable accents the divergent paths of groups and classes. These paths do not find their explanation in the self-interested decisions of interacting individuals. They grow out of the deployment of social labor, mobilized to engage the world of nature. The manner of that mobilization sets the terms of history, and in these terms the peoples who have asserted a privileged relation with history and the peoples to whom history has been denied encounter a common destiny.

Thus, globalist notions of historical development that have elevated particular interests and visions to the level of universal principles must be rejected (see Mikecin 1995). The idea of a single world economic system based upon a free market is as utopian as that of a single set of values or a single type of economic, political, or sociocultural life. In reality, the world is characterized by plurality. Different cultures and societies develop different types of economic and social lives; no single type can pretend to be universal or modern. Hence, there are many paths to modernity.

According to the worldview held by various contemporary political elites, economic efficiency is distinct from human welfare. This distinction implies an instrumentalization of human beings, in the sense that people serve markets and not the other way around. The processes of globalization, as Gray (2002) points out, are aimed at legitimizing such aspirations. Should these depersonalized attitudes prevail within global capitalism, its culture of consumerism will be viewed as an unsustainable one. However, some societies might ultimately embrace globalization, albeit critically and prudently, making the most of what it has to offer and attempting to minimize the consequences of, or altogether avoid, what Lay (2003) sees as the unfavorable processes and solutions that may be harmful to cultural, economic, and environmental sustainability.

Globalization, understood as the spread of new technologies that eliminate distances, does not universalize Western values; on the contrary, it makes the world pluralistic. The existing interdependence among the economies of the world, therefore, does not entail a unified, economic culture. Anthropologists studying globalization and sustainability face the task of documenting and understanding the *modus vivendi* among societies within this world of nations, brought about by global processes yet yielding diverse societal outcomes. Milton (1997: 492) states that anthropology, in particular, can help us understand "what a sustainable way of living might entail, not only in terms of how the environment is physically treated, but in terms of what kinds of values, beliefs, kinship structures, political ideologies and ritual traditions might support sustainable practices." Therefore, if anthropology is to critically analyze sustainability in the contemporary world, it will have to approach the study of culture through multiple perspectives, based upon the interests and needs of particular societies, rather than the universalist interests of any single dominant ideological, methodological, or historical tradition.

Translated from Croatian by Nataša Pavlović

Notes

1. Consumption within the bounds of the world system is always a consumption of identity canalized by a negotiation between self-definition and the array of possibilities offered by the capitalist market. See Friedman (1990: 314).

2. As an illustration, I will mention an example from my own experience. When I was in Mexico in 1982 on a Mexican government grant, I wanted to take part in the 10th World Congress of Sociology. On the second day of the Congress I came to the venue but, instead of the workshop I was interested in, there was a group of UNAMA (Universidad Nacional Autónoma de México) students who had staged a protest. A banner was displayed saying. "Down with colonialism in social sciences!" When I asked them what they meant by the slogan, one of them told me that they were protesting against English being the only official language of the Congress, and demanding that Spanish should be one as well. I replied jokingly that what they were asking for meant only adding another colonial language to the one already in use. If they wanted to be consistent, I said, they should also demand the use of one or several Indian languages, such as Nahuatl or one of the Mayan languages. The protester gave me a blank look, as he had obviously not viewed the issue from that angle. What is at issue here is the discourse of "cultural authenticity," because of the problem of deciding between the claims of a "Latin" culture organized around the nation-state built by the original European colonizers and an indigenous "American" one. Once the language of the colonizers, Spanish has become naturalized in that culture, while English is still seen as colonial.

3. It is possible for people to "imagine the community" of the nation-state, even though this community is spatially extended. But this is not possible on the global level since there are neither effective global institutions regulating practices at this level nor any cultural representations of "global identity" (See Tomlinson 1992: 175).

4. It is therefore useful to distinguish between the countries that are commanding globalization–the globalizers (wealthy nations)–and the countries that suffer globalization, the globalized countries (poor nations). See Gadotti (2003: 6-7).

5. One such model would be that of an ecosocial market economy, coupling a competitive (efficiency- and ownership-oriented) market economy with social justice (solidarity) built on a legislative framework. Ecology, society, and market economy are the cornerstones of a strategic triangle and at the same time form the foundation of the ecosocial market economy, which blazes the path to a sustainable future. See Radermacher (2003: 12).

References

Ang, Ien, ed. 1996. *Living Room Wars: Rethinking Media Audiences for a Postmodern World.* London: Routledge.

Bodley, John H. 1999. *Victims of Progress* 4th ed. London: Mayfield.

___. 2001. *Anthropology and Contemporary Human Problems,* 4th ed. London: Mayfield.

Cifrić, Ivan. 2001. "Eskurs o održivom razvoju." *Socijalna Ekologija* 10 (3): 157-69.

Čolić, Snježana. 2002. *Kultura i povijest.* Zagreb: Hrvatska sveučilišna naklada.

Daly, Herman E., and John B. Cobb, Jr. 1989. *For the Common Good: Redirecting the Economy Toward Community, the Environment, and a Sustainable Future.* Boston: Beacon Press.

Eder, Klaus. 1990. "The Cultural Code of Modernity and the Problem of Nature: A Critique of the Naturalistic Notion of Progress." In *Rethinking Progress,* ed. Jeffrey C. Alexander and Piotr Sztompka. London and Boston: Unwin Hyman.

Featherstone, Mike. 1991. *Consumer Culture and Postmodernism.* London: Sage.

Fox, Jeremy. 2001. *Chomsky i globalizacija.* Zagreb: Jesenski i Turk.

Friedman, Jonathan. 1990. "Being in the World: Globalization and Localization." *Theory, Culture & Society* 7: 311-28.

Gadotti, Moacir. 2003. "Pedagogy of the Earth and Culture of Sustainability." Paper presented at *Lifelong Citizenship Learning, Participatory Democracy & Social Change,* October 17-19, Transformative Learning Centre, Ontario Institute for Studies in Education, University of Toronto.

Gray, John. 2002. *Lažna zora.* Zagreb: Masmedia.

Harvey, David. 1989. *The Condition of Postmodernity: An Inquiry into the Origins of Cultural Change.* Oxford: Blackwell.

Jacques, Martin. 1989. "Britain and Europe." In *New Times: The Changing Face of Politics in the 1990s,* ed. Stuart Hall and Martin Jacques. London: Lawrence and Wishart.

Jameson, Fredric. 1984. "The Politics of Theory: Ideological Positions in the Postmodernism Debate." *New German Critique* 33 (Fall): 53-65.

Kalanj, Rade. 1994. *Modernost i napredak.* Zagreb: Aktant.

Küng, Hans. 1997. *Weltethos für Weltpolitik und Weltwirtschaft.* Munich and Zurich: Piper.

Lay, Vladimir. 1992. "Održivi razvitak i društvene promjene." *Socijalana ekologija* 1 (1): 1-17.

___. 2003. "Proizvodnja budućnosti Hrvatske: integralna održivost kao koncept i kriterij." *Društvena istraživanja* 12 (3-4): 311-34.

Mikecin, Vjekoslav. 1995. *Umjetnost i povijesni svijet.* Zagreb: Hrvatsko Filozofsko Drustvo.

Pearce, David W., and Jeremy J. Warford. 1993. *World Without End: Economics Environment and Sustainable Development.* New York: Oxford University Press.

Milton, Kay. 1997. "Ecologies: Anthropology, Culture and the Environment." *International Social Science Journal* 154: 477-95.

Pieterse, Nederveen Jan. 1995. "Globalisation as Hybridisation." *International Sociology* 9 (2): 161-84.

Radermacher, Franz Josf. 2003. *Ravnoteza ili razaranje.* Zagreb: Intercon Nakladni zavod Globus.

Robertson, Roland. 1990. "Mapping the Global Condition: Globalization as the Central Concept." *Theory, Culture & Society* 7 (2-3): 15-30.

Schmookler, Andrew Bard. 1991. "The Insatiable Society: Materialistic Values and Human Needs." *The Futurist,* July-August, 17-19.

Sklair, Leslie. 1991. *Sociology of the Global System.* Baltimore: The Johns Hopkins University Press.

Smith, Anthony D. 1990. "Towards a Global Culture." *Theory, Culture & Society* 7 (2-3): 171-91.

Stavenhagen, Rudolfo. 1976. "Kako izvrsiti dekolonizaciju primenjenih drustvenih nauka." *Marksizam u svetu* 6: 138-47.

Storey, John. 2003. *Inventing Popular Culture.* London: Blackwell.

Tomlinson, John. 1992. *Cultural Imperialism.* Baltimore: The Johns Hopkins University Press.

Urry, John. 1989. "The End of Organised Capitalism." In *The Times: The Changing Face of Politics in the 1990s,* ed. Stuart Hall and Martin Jacques. London: Lawrence and Wishart.

Wallerstein, Immanuel. 1986. *Suvremeni svjetski sistem.* Zagreb: Cekade.

___. 1990. "Culture as the Ideological Battleground of the Modern World-System." *Theory, Culture & Society* 7 (2-3): 31-55.

Wolf, Eric. 1982. *Europe and People Without History.* Berkeley: University of California Press.

World Commission on Environment and Development (WCED). 1987. *Our Common Future.* Oxford: Oxford University Press.

Appendix 1: Glossary

Sustainability:

"improving the quality of human life while living within the carrying capacity of supporting eco-systems." – The World Conservation Union, United Nations Environment Program, World Wide Fund for Nature. 1991. *Caring for the Earth: A Strategy for Sustainable Living.* Gland, Switzerland.

"Sustainability encompasses the simple principle of taking from the earth only what it can provide indefinitely, thus leaving future generations no less than we have access to ourselves. – "Friends of the Earth Scotland

Sustainable development:

"Sustainable development is development that meets the needs of the present without compromising the ability of future generations to meet their own needs." – The Brundtland Report, World Commission on Environment and Development. 1987. *Our Common Future.* Oxford: Oxford University Press, p. 8.

Sustainable Communities:

"In a sustainable community, resource consumption is balanced by resources assimilated by the ecosystem. The sustainability of a community is largely determined by the web of resources providing its food, fiber, water, and energy needs and by the ability of natural systems to process its wastes. A community is unsustainable if it consumes resources faster than they can be renewed, produces more wastes than natural systems can process or relies upon distant sources for its basic needs." – Sustainable Community Roundtable Report (South Puget Sound)

"Sustainable communities foster commitment to place, promote vitality, build resilience to stress, act as stewards, and forge connections beyond the community." – Northwest Policy Institute (University of Washington Graduate School of Public Affairs, Seattle, Washington

Communities of place:

"They are complex, human-scaled places that combine many elements of living: public, private work, and home. They mix different kinds of people and activities in close proximity and provide places for them to interact. They provide for everyday and sometimes random casual meetings that foster a sense of community. They create shared places that are unique to each neighborhood and shape a social geography intimately known only by those who live or work there. They are hard to design but easy to design away. And they are essential to our well-being—not just in times of crisis, but also in living our everyday lives." – Peter Calthorpe and William Fulton. 2001. *The Regional City: Planning for the End of Sprawl.* Washington, DC: Island Press, p. 31.

Notes on Contributors

JANET E. BENSON is an associate professor of anthropology in the Department of Sociology, Anthropology, and Social Work at Kansas State University. She is past president of the Committee on Refugees and Immigrants (CORI) of the American Anthropological Association. Her research focuses on meatpacking and immigration in southwest Kansas.

KARLA CASER teaches in the Architecture and Urbanism Studio at Faculdade Nacional (FINAC) and at Centro Federal de Educação Tecnológica do Espírito Santo (CEFETES), both located in Vitória, Brazil. She is a professional architect and urbanist and the principal of her own firm. Her research focuses on the use of sustainable building technologies and on human-environment relationships in design practice.

SNJEŽANA ČOLIĆ is a research associate in the Institute of Social Sciences "Ivo Pilar" in Zagreb, Croatia, and an associate professor in the sociology of culture at the Center for Croatian Studies, University of Zagreb. Her research focuses on localization, globalization, and the role of social and cultural capital in Croatian development.

PAULO DA CUNHA LANA is an associate professor and head of the Postgraduate Program in Environment and Development (UNESCO Chair on Sustainable Development) of the Universidade Federal do Paraná, Brazil. His research focuses on coastal ecosystems, sustainable development, and biodiversity in Brazil.

ANGELA DUARTE D. FERREIRA is a professor in the Postgraduate Programs in Environment and Development and in sociology at the Universidade Federal do Paraná, Brazil. Her research focuses on rural sociology, agriculture and development, and agrarian reform.

JOHANNA GIBSON is a lecturer in intellectual property law, Queen Mary Intellectual Property Research Institute, within the Centre for Commercial Law Studies, University of London. She is project director of "Patenting Lives," an international network examining cultural and socioeconomic aspects of patents on life forms. Her research focuses on international property and development, traditional knowledge, community governance, and cultural diversity.

KRISTA HARPER is an assistant professor in the Department of Anthropology and the Center for Public Policy and Administration (CPPA) at the University of Massachusetts, Amherst. She received a European Union Affairs Research Fellowship through the Fulbright Scholar Program to study ethnicity and Roma civil rights. Her research focuses on social movements and the politics of environment and health.

CARL A. MAIDA is a professor of public health at the University of California, Los Angeles. He is president of the American Association for the Advancement of Science, Pacific Division. His research focuses on health and the environment in immigrant communities, ethnic cultural dimensions of HIV care, within the UCLA AIDS Institute, and natural hazards, within the National Center for Child Traumatic Stress at the David Geffen School of Medicine.

BARBARA YABLON Maida teaches in the Department of Geography at the University of California, Los Angeles. Her research focuses on medical topologies, with emphasis on the North American agricultural frontier and its relationship to the contemporary reemergence of arboviruses within the irrigated California Central Valley.

KENNETH A. METER is principal at the Crossroads Resource Center in Minneapolis and teaches in the Department of Applied Economics at the University of Minnesota. He serves as technical consultant for the City of Minneapolis Sustainability Initiative.

DARIO NOVELLINO is an honorary research fellow at the University of Kent at Canterbury, UK. He is a visiting research associate of the Institute of Philippine Culture of the Ateneo de Manila University and has carried out studies for the International Fund for Agricultural Development of the United Nations in Indonesia and Vietnam. His research focuses on ethnobotany, natural resource management, conservation, and indigenous rights.

DEBORAH PELLOW is a professor in the Department of Anthropology at Syracuse University and is co-director of the Space and Place Initiative in the Global Affairs Institute at the Maxwell School. She was a Fulbright Teaching Fellow at

Osaka University and Ritsumeikan University in Japan. Her research focuses on urban spatial forms, social organization, and community life in West Africa.

CLAUDE RAYNAUT is a research director at the Centre National de la Recherche Scientifique (France), a professor at the Université de Bordeaux 2, and an associate professor at the Universidade Federal do Paraná, Brazil. His research in several African countries and in Brazil has focused on the interactions between local populations and their natural habitat, poverty and child health, and social dimensions of the AIDS epidemic.

THOMAS F. THORNTON is a visiting associate professor of anthropology at Trinity College in Hartford, Connecticut. His research focuses on Tlingit and Alaska Native history, geography, and subsistence issues.

RICHARD WESTRA is an assistant professor in the Division of International and Area Studies, Pukyong National University, Pusan, South Korea. His research focuses on the political economy of South Korean development, Marxist theory and social change, and globalization and the state.

MAGDA ZANONI is a professor at the Universidade Federal do Paraná, Brazil, a professor in the Postgraduate Program in Rural Development at Université Fédérale du Rio Grande do Sul, Brazil, and a lecturer at the Université Paris 7 – Denis Diderot. Her research focuses on environment and development, society-nature relations, and biodiversity conservation in Brazil and rural France.

Index

Lightning Source UK Ltd.
Milton Keynes UK
UKOW041350250113

205320UK00009B/374/P